Monographs in Theoretical Computer Science
An EATCS Series

Springer
Berlin
Heidelberg
New York
Barcelona
Budapest
Hong Kong
London
Milan
Paris
Santa Clara
Singapore
Tokyo

Kurt Jensen

Coloured Petri Nets

Basic Concepts, Analysis Methods and Practical Use

Volume 2

With 34 Figures

Author

Prof. Kurt Jensen
Aarhus University
Computer Science Department
Ny Munkegade, Bldg. 540
DK-8000 Aarhus C, Denmark

Series Editors

Prof. Dr. Wilfried Brauer
Institut für Informatik, Technische Universität München
Arcisstrasse 21, D-80333 München, Germany

Prof. Dr. Grzegorz Rozenberg
Department of Computer Science
University of Leiden, Niels Bohrweg 1, P.O. Box 9512
2300 RA Leiden, The Netherlands

Prof. Dr. Arto Salomaa
Data City
Turku Centre for Computer Science
FIN-20520 Turku, Finland

Cataloging-in-Publication Data applied for
Die Deutsche Bibliothek - CIP-Einheitsaufnahme

Jensen, Kurt:
Coloured petri nets: basic concepts, analysis methods and practical use / Kurt Jensen. - Berlin; Heidelberg; New York; Barcelona; Budapest; Hong Kong; London; Milan; Paris; Santa Clara; Singapore; Tokyo: Springer
(Monographs in theoretical computer science)
Vol. 2.. - 2., corr. printing. - 1997

Second corrected printing 1997

ISBN 978-3-642-08200-9

Cover Design: MetaDesign, Berlin

Note to the Corrected Reprint

This corrected reprint is identical to the first edition - except for the correction of a few errors. Most of these are simple spelling mistakes, duplicated words, and other small errors that have little or no influence on the readability of the text.

Despite all efforts some errors remain. That seems to be inevitable, no matter how many people read the manuscript. If you wish to report errors or discuss other matters you may contact me via electronic mail: kjensen@daimi.aau.dk. You may also take a look at my WWW pages: http://www.daimi.aau.dk/~kjensen/. They contain a lot of material about CP-nets and the CPN tools, including a list of errata for this book.

Aarhus, Denmark
February 1997

Kurt Jensen

Preface

This volume contains a detailed presentation of the analysis methods for CP-nets. The analysis methods allow the modeller to investigate dynamic properties of CP-nets, e.g., the properties defined in Chap. 4 of Vol. 1. We describe the main ideas behind the analysis methods and the mathematics on which they are based. We also describe how the methods are used in practice and how they are supported by computer tools.

Some parts of this volume are rather theoretical while other parts are application oriented. The purpose of the volume is to teach the reader how to use the formal analysis methods. This does not require a deep understanding of the underlying mathematical theory (although such knowledge will, of course, be a help). Chapters 1–3 deal with occurrence graphs, Chap. 4 deals with invariants, and Chap. 5 defines the behaviour of timed CP-nets.

In Chap. 5 of Vol. 1 we introduced the basic ideas behind occurrence graphs and place invariants. The descriptions there were intuitive, imprecise and without proofs. In this volume we give the precise unambiguous definitions, prove the correctness of the various techniques, and explain how to use the methods in practice. To make the description complete we repeat the informal introductions provided in Vol. 1.

How to read Volume 2

In this volume we assume that the reader is familiar with the basic concepts of CP-nets, in particular the formal definition (in Sects. 2.2–2.3 of Vol. 1) and the dynamic properties (in Sects. 4.2–4.5 of Vol. 1). The analysis methods are defined in terms of markings, steps, token elements, and binding elements. This means they can be used both for hierarchical and for non-hierarchical CP-nets. It is possible to read and understand most of the material in this volume without knowing the formal definition of hierarchical CP-nets.

The volume contains three parts. They deal with occurrence graphs, place/transition invariants, and timed CP-nets, respectively. The three parts are independent of each other, and hence they can be read in any order. The only exception to this is Sect. 5.6, which discusses how timed CP-nets can be analysed by means of occurrence graphs and place/transition invariants.

Readers who are interested in both the theory and practical use of CP-nets are advised to read all chapters (in the order in which they appear). Readers who are primarily interested in the practical use of CP-nets may find it sufficient to read the sections that contain introductions and examples and skip the sections containing the formal definitions and proofs. Readers who are primarily interested in the theoretical aspects of CP-nets may decide to skip the sections dealing with tool support.

All readers are advised to do the exercises of the chapters they read (or at least some of them). It is also recommended to have access to the CPN tools described in the various chapters – or access to other Petri net tools.

At Aarhus University, Vol. 2 is being used as the material for a graduate course. The course has 10–12 lectures of 2 hours each. This is supplemented by 8–10 classes discussing the exercises. The students also undertake a small project in which they either use the analysis methods on some of their CPN models or do some more theoretical work, e.g., related to one of the analysis methods. The students are expected to use one third of their study time on this course, during 4 months.

Acknowledgements

Many different people have contributed to the development of the analysis methods and tools for CP-nets. Below some of the most important contributions are listed.

The first version of occurrence graphs with equivalence classes was developed together with Peter Huber, Arne Møller Jensen and Leif Obel Jepsen. Later work has been done together with Rikke Drewsen Andersen, Søren Christensen, Afshin Foroughipour, Tommy Rudmose Hansen, Peter Huber, Jens Bæk Jørgensen and Michael Pedersen.

The CPN tool for occurrence graphs was designed together with Søren Christensen and Peter Huber. The implementation was done by Peter Huber. The CPN tool for place invariants was designed together with Søren Christensen. The implementation was done by Søren Christensen and Jan Toksvig.

Meta Software provided the financial support for the CPN tool project. So far, more than 35 man-years have been used. The project is also supported by the Danish National Science Research Council, the Human Engineering Division of the Armstrong Aerospace Medical Research Laboratory at Wright-Patterson Air Force Base, and the Basic Research Group of the Technical Panel C3 of the US Department of Defense Joint Directors of Laboratories at the Naval Ocean Systems Center.

In addition to those mentioned above, a number of students and colleagues have read and commented on earlier versions of this volume. In particular, I am grateful to Jonathan Billington, Torben Bisgaard Haagh, Niels Damgaard Hansen, Charles Lakos, Kjeld Høyer Mortensen, Laure Petrucci, Dan Simpson and Antti Valmari (scientific and linguistic assistance) – and to Karen Kjær Møller and Andrew Ross (linguistic assistance).

Despite all help some errors remain. That seems to be inevitable, no matter how many people read the manuscript. If you wish to report errors or discuss other matters you may contact me via electronic mail: kjensen@daimi.aau.dk.

Aarhus, Denmark
July 1994

Kurt Jensen

Table of Contents

Chapter 1

Full Occurrence Graphs

This chapter deals with the most basic kind of occurrence graphs – called full occurrence graphs (O-graphs). It is relatively straightforward to define, understand, construct, and analyse O-graphs. They can be used to verify all the dynamic properties defined in Vol. 1, i.e., reachability, boundedness, home, liveness, and fairness properties. The construction of O-graphs and the associated verification of dynamic properties can be fully automated. This means O-graphs provide a very straightforward and easy-to-use method to analyse the properties of a given CP-net.

The occurrence graphs in this chapter (and in Chaps. 2–3) are defined for hierarchical CP-nets. However, each non-hierarchical CP-net can be considered to be a hierarchical CP-net with only one page (and one page instance). Thus we can also use the concepts for non-hierarchical nets. Then the proof rules speak about places (instead of place instances) and transitions (instead of transition instances).

Section 1.1 contains an informal introduction to O-graphs. This is done by means of an example. Section 1.2 contains the formal definition of O-graphs. Section 1.3 defines a number of standard concepts from graph theory, in particular directed paths and strongly connected components. Section 1.4 contains a number of detailed proof rules, i.e., propositions that allow us to prove (or disprove) different CP-net properties by inspection of different O-graph properties. Sections 1.5 and 1.6 contain O-graphs for two different CP-nets modelling a distributed data base system and a set of dining philosophers. These sections show how fast the sizes of O-graphs may grow when the sizes of the involved colour sets increase. Finally, Sect. 1.7 discusses how the construction and analysis of O-graphs are supported by computer tools.

As mentioned above, O-graphs are easy to construct and analyse. However, they are often too big to be used in practice. In Chaps. 2 and 3 we shall introduce other kinds of occurrence graphs. They are more complex to understand and use, but often much smaller.

1.1 Introduction to O-graphs

The basic idea behind occurrence graphs is to construct a graph which has a node for each reachable marking and an arc for each occurring binding element. Obviously, such a graph may become very large, even for small CP-nets. As an example, let us consider the resource allocation system from Fig. 1.7 of Vol. 1. Due to the cycle counters this net has an infinite number of reachable markings and thus an infinite occurrence graph. However, we can simplify the CP-net by omitting the cycle counters. Then we get the CP-net shown in Fig. 1.7. It is easy to check that the cycle counters form an isolated part of the original CP-net – in the sense that they influence neither the enabling nor the effect of an occurrence (except that they determine the values of new cycle counters). This means the simplified net in Fig. 1.1 has a behaviour similar to that of the original net in Fig. 1.7 of Vol. 1. For each occurrence sequence in one of the CP-nets there is a corresponding occurrence sequence in the other. Hence we can get information about the dynamic properties of the original net by constructing an occurrence graph for the simplified net.

Such a graph is shown in Fig. 1.2 – it is called a **full occurrence graph** or an **O-graph**. Each node in an O-graph represents a reachable marking, and the content of this marking is described by the text inscription of the node. The left

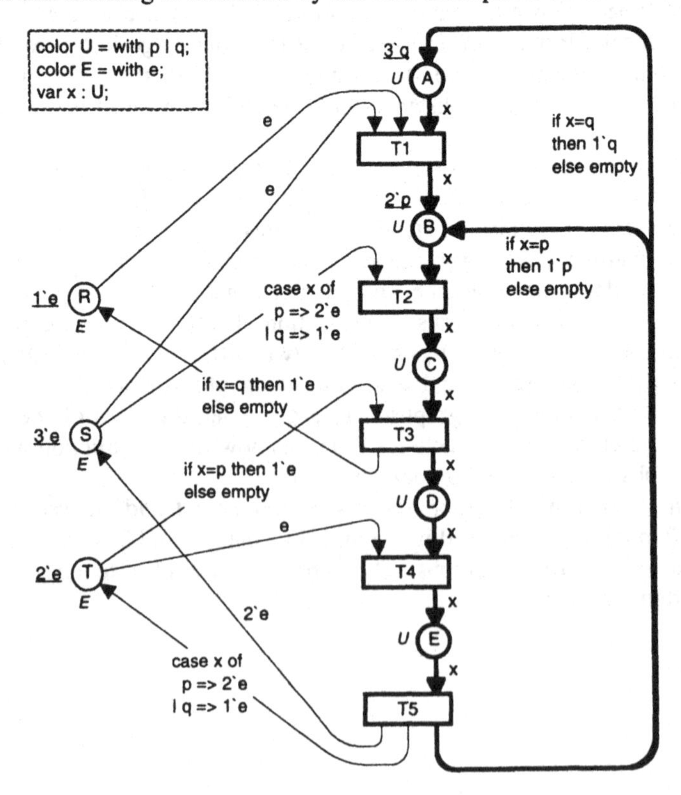

Fig. 1.1. Simplified CP-net describing the resource allocation system

column contains the number of e-tokens on the three resource places R–T, while the right column contains the p-tokens and q-tokens on the places A–E (to improve readability we use P and Q, instead of p and q). The node with a thick border line represents the initial marking. We often give each node a unique node number. This makes it easy to refer to the individual nodes. The node numbers have no semantic meaning. Each arc in an O-graph represents the occurrence of a binding element, and the content of this binding element is described by the text attached to the arc. The left component of the pair indicates the transition, while the right one indicates the value bound to x. There is an arc, with binding element b, from a marking M_1 to a marking M_2 iff M_1 [b> M_2.

Notice that in occurrence graphs we usually omit arcs that correspond to steps containing more than one binding element. Otherwise, we would have had, e.g., an arc from node #1 to node #6, with the inscription 1\`(T2,P)+1\`(T1,Q). Such arcs would give us information about the concurrency between binding elements, but they are not necessary for the verification of the dynamic properties defined in Chap. 4 of Vol. 1.

From the O-graph in Fig. 1.2 it is possible to verify all the dynamic properties postulated for the resource allocation system in Chap. 4 of Vol. 1 (see Exercise 1.1).

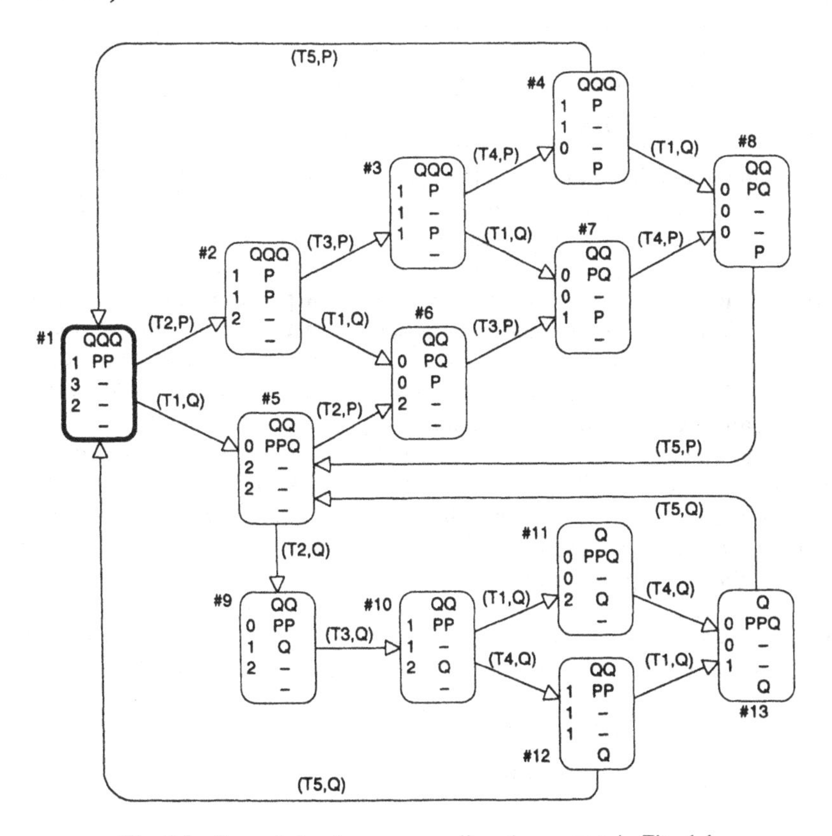

Fig. 1.2. O-graph for the resource allocation system in Fig. 1.1

Even for a small O-graph, like the one in Fig. 1.2, the construction and investigation are tedious and error-prone. In practice, it is not unusual to handle CP-nets that have O-graphs containing more than 100,000 nodes (and many CP-nets have millions of markings). Thus it is obvious that we need to be able to construct and investigate the O-graphs by means of a computer, and also that we want to develop techniques by which we can construct reduced occurrence graphs without losing too much information. As shown in Chaps. 2 and 3, there are several different ways to obtain such a reduction. For all of them, it is important to notice that the reduced occurrence graph can be obtained directly, i.e., calculated without first constructing the O-graph.

1.2 Formal Definition of O-graphs

In this section we define the O-graph of a CP-net, and we also give an algorithm to construct it. First we recapture one of the basic definitions from graph theory:

Definition 1.1: A **directed graph** is a tuple DG = (V, A, N) such that:

(i) V is a set of **nodes** (or **vertices**).
(ii) A is a set of **arcs** (or **edges**) such that:
 • $V \cap A = \emptyset$.
(iii) N is a **node** function. It is defined from A into $V \times V$.

DG is **finite** iff V and A are finite.

It should be noted that, in contrast to classical graph theory, we allow a directed graph to have several arcs between the same ordered pair of nodes (and thus we define A as a separate set and not as a subset of $V \times V$).

An arc a with $N(a) = (v_1, v_2)$ is said to go from the **source node** v_1 to the **destination node** v_2. We also say that a is an **output arc** of v_1 and an **input arc** of v_2. A node with no output arcs is said to be **terminal**, and we use V_T to denote the set of all terminal nodes. Finally, we define two functions $s, d \in [A \rightarrow V]$. The first function maps each arc into its source node, while the second maps each arc into its destination node.

Two directed graphs are said to be isomorphic iff they have the same structure, i.e., when it is possible to define a one-to-one mapping between the two sets of nodes and a one-to-one mapping between the two sets of arcs, in such a way that each arc connects those two nodes which correspond to the source and destination of the corresponding arc:

Definition 1.2: The directed graph DG* = (V*, A*, N*) is **isomorphic** to DG iff there exist two bijections $B_V \in [V \rightarrow V^*]$ and $B_A \in [A \rightarrow A^*]$ such that:

(i) $\forall a \in A$: $(B_V(s(a)), B_V(d(a))) = (s(B_A(a)), d(B_A(a)))$.

Now we are ready to define full occurrence graphs:

Definition 1.3: The **full occurrence graph** of a CP-net, also called the **O-graph**, is the directed graph $OG = (V, A, N)$ where:

(i) $V = [M_0\rangle$.
(ii) $A = \{(M_1,b,M_2) \in V \times BE \times V \mid M_1 [b\rangle M_2\}$.
(iii) $\forall a=(M_1,b,M_2) \in A$: $N(a) = (M_1,M_2)$.

The O-graph has a node for each reachable marking and an arc for each step that occurs – with a single binding element. The source node of the arc is the start marking of the step, while the destination node is the end marking.

When we draw O-graphs, like the one in Fig. 1.2, we usually inscribe each node with a text string describing the marking which the node represents. To save space, we sometimes use a condensed representation of the marking. Analogously, we inscribe each arc with the binding element which it represents. For an arc (M_1,b,M_2), it would be redundant to include the two markings M_1 and M_2 in the arc inscription, because these two markings are already described via the node inscriptions of the source node and the destination node.

When we have a CP-net where all variables (in the arc expressions and guards) have finite colour sets, it is straightforward to prove that the O-graph is finite iff all place instances are bounded.

Below we give an abstract algorithm to construct the O-graph. *Waiting* is a set of nodes. It contains those nodes for which we have not yet found the successors. *Node(M)* is a procedure that creates a new node M and adds M to *Waiting*. If M is already a node, the procedure has no effect. Analogously, *Arc(M_1,b,M_2)* is a procedure that creates a new arc (M_1,b,M_2) with source M_1 and destination M_2. If (M_1,b,M_2) is already an arc, the procedure has no effect (this never happens for O-graphs but it may happen for the OE-graphs and OS-graphs which we introduce in Chaps. 2 and 3). For a marking $M_1 \in \mathbb{M}$ we use $Next(M_1)$ to denote the set of all possible "next moves":

$$Next(M_1) = \{(b,M_2) \in BE \times \mathbb{M} \mid M_1 [b\rangle M_2\}.$$

Proposition 1.4: The following algorithm constructs the O-graph. The algorithm halts iff the O-graph is finite. Otherwise the algorithm continues forever, producing a larger and larger subgraph of the O-graph.

```
Waiting := Ø
Node(M0)
repeat
    select a node M1 ∈ Waiting
    for all (b,M2) ∈ Next(M1) do
    begin
        Node(M2)
        Arc(M1,b,M2)
    end
    Waiting := Waiting – {M1}
until Waiting = Ø.
```

Proof: Straightforward consequence of Def. 1.3. □

When the O-graph is infinite or too big to be constructed by the available computing power, it may still be of interest to construct a partial O-graph, i.e., a subgraph of the O-graph. We shall return to this in Sect. 1.7 when we discuss computer support of the occurrence graph method.

1.3 Directed Paths and Strongly Connected Components

This section defines a number of standard concepts from graph theory, in particular directed paths, strongly connected components and SCC-graphs. All definitions are given with respect to a directed graph DG = (V, A, N).

Definition 1.5: A **finite directed path** is a sequence of nodes and arcs:

$$v_1\ a_1\ v_2\ a_2\ v_3 \ldots v_n\ a_n\ v_{n+1}$$

such that $n \in \mathbb{N}$, and $N(a_i) = (v_i, v_{i+1})$ for all $i \in 1..n$. The node v_1 is called the **start node** of the path, while the node v_{n+1} is called the **end node**. The non-negative integer n is the **length** of the path.

Analogously, an **infinite directed path** is a sequence of nodes and arcs:

$$v_1\ a_1\ v_2\ a_2\ v_3 \ldots$$

such that $N(a_i) = (v_i, v_{i+1})$ for all $i \in \mathbb{N}_+$. The node v_1 is called the **start node** of the path, which is said to have **infinite length**.

The set of all finite directed paths is denoted by DPF, and we use $DPF(v_1,v_2)$ to denote the set of all finite directed paths which start in v_1 and end in v_2. The set of all infinite directed paths is denoted by DPI. Finally, we use $DP = DPF \cup DPI$ to denote the set of all directed paths.

A finite directed path is a **directed cycle** iff the start node and the end node are identical and the length is positive. A directed cycle is **simple** iff all nodes in it are different (except that $v_1 = v_n$). A directed graph is **acyclic** iff it has no directed cycles. The set of all simple directed cycles is denoted by DCS.

For a node $v \in V$ and a directed path $dp \in DP$ we use the notation $v \in dp$ to denote that v is one of the nodes in dp. A similar notation is used for arcs.

The way we have defined directed paths of a directed graph (in Def. 1.5) is analogous to the way we defined occurrence sequences of a CP-net (in Def. 2.10 of Vol. 1). This means there is a close relationship between the directed paths of the O-graph and the occurrence sequences of the CP-net:

Proposition 1.6: The O-graph satisfies the following properties:

(i) Each finite occurrence sequence:

$$M_1 [b_1\rangle M_2 [b_2\rangle M_3 \ldots M_n [b_n\rangle M_{n+1}$$

where $M_1 \in [M_0\rangle$ and $b_i \in BE$ for all $i \in 1..n$, has a **matching directed path**:

$$M_1 \ (M_1,b_1,M_2) \ M_2 \ (M_2,b_2,M_3) \ M_3 \ldots M_n \ (M_n,b_n,M_{n+1}) \ M_{n+1}.$$

(ii) Each finite directed path:

$$M_1 \ (M_1,b_1,M_2) \ M_2 \ (M_2,b_2,M_3) \ M_3 \ldots M_n \ (M_n,b_n,M_{n+1}) \ M_{n+1}$$

has a **matching occurrence sequence**:

$$M_1 [b_1\rangle M_2 [b_2\rangle M_3 \ldots M_n [b_n\rangle M_{n+1}.$$

Similar properties are satisfied for infinite occurrence sequences and infinite directed paths.

Proof: The two properties are proved by means of induction over the length of the occurrence sequence/directed path.

Property (i): When the length of the occurrence sequence is zero, property (i) follows directly from Def. 1.3 (i). Next, assume that property (i) is satisfied when the length of the occurrence sequence is n–1, and assume that we have a finite occurrence sequence:

$$M_1 [b_1\rangle M_2 [b_2\rangle M_3 \ldots M_n [b_n\rangle M_{n+1}$$

where $M_1 \in [M_0\rangle$ and $b_i \in BE$ for all $i \in 1..n$. From the inductive hypothesis, it follows that there exists a finite directed path:

$$M_1 \ (M_1,b_1,M_2) \ M_2 \ (M_2,b_2,M_3) \ M_3 \ldots M_{n-1} \ (M_{n-1},b_{n-1},M_n) \ M_n.$$

From the above occurrence sequence, we see that M_{n+1} is reachable from $M_1 \in [M_0\rangle$. Hence M_{n+1} is reachable from M_0, and from Def. 1.3 (i) we know that M_{n+1} belongs to V. From Def. 1.3 (ii) and the step $M_n [b_n\rangle M_{n+1}$ we conclude that (M_n,b_n,M_{n+1}) is an arc of the O-graph. This means that:

$$M_1 \ (M_1,b_1,M_2) \ M_2 \ (M_2,b_2,M_3) \ M_3 \ldots M_n \ (M_n,b_n,M_{n+1}) \ M_{n+1}$$

is a finite directed path, as required.

Property (ii): When the length of the directed path is zero, property (ii) is trivially satisfied. Next, assume that property (ii) is satisfied when the length of the directed path is n–1, and assume that we have a finite directed path:

$$M_1 \ (M_1,b_1,M_2) \ M_2 \ (M_2,b_2,M_3) \ M_3 \ldots M_n \ (M_n,b_n,M_{n+1}) \ M_{n+1}.$$

From the inductive hypothesis, it follows that there exists a finite occurrence sequence:

$$M_1 [b_1\rangle M_2 [b_2\rangle M_3 \ldots M_{n-1} [b_{n-1}\rangle M_n.$$

From Def. 1.3 (ii) and the arc (M_n, b_n, M_{n+1}) we conclude that $M_n [b_n\rangle M_{n+1}$ is a step of the CP-net. This means that:

$$M_1 [b_1\rangle M_2 [b_2\rangle M_3 \ldots M_n [b_n\rangle M_{n+1}$$

is a finite occurrence sequence, as required. □

In the definition of O-graphs we have only considered steps containing a single binding element. This means each directed path corresponds to an occurrence sequence where each step has a single binding element, instead of a *multi-set* of binding elements. However, when an enabled step has more than one binding element, we know that these binding elements can occur in any order and that the order does not influence the effect. Hence we can always transform an occurrence sequence with sets of binding elements into an occurrence sequence with single binding elements.

It is easy to verify that the modified occurrence sequence has the same behavioural properties as the original occurrence sequence – as long as we do not consider the question of concurrency. Hence we can investigate all the dynamic properties introduced in Chap. 4 of Vol. 1, by means of occurrence graphs where each arc represents a single binding element. If we want to use occurrence graphs to investigate whether binding elements are concurrent or not, we need to construct occurrence graphs where each possible step is represented by an arc, instead of a *sequence* of arcs. Such a graph has the same number of nodes as the corresponding ordinary occurrence graph, but it may have many more arcs.

In the rest of this book we only consider occurrence graphs where each arc represents a single binding element (or an equivalence class of single binding elements; see Chaps. 2 and 3).

Definition 1.7: The directed graph $DG^* = (V^*, A^*, N^*)$ is a **subgraph** of DG iff the following properties are satisfied:

(i) $V^* \subseteq V$.
(ii) $A^* \subseteq \{a \in A \mid N(a) \in V^* \times V^*\}$.
(iii) $N^* = N \mid A^*$.

When the inclusion sign of (ii) can be replaced by an equality sign, DG* is said to be the subgraph **induced** by V*.

The above definitions of subgraphs and induced subgraphs are analogous to the corresponding definitions for bipartite directed graphs and non-hierarchical CP-nets given in Sect. 4.1 of Vol. 1.

Next we consider strongly connected sets and strongly connected components. A set of nodes V* is strongly connected iff, for any pair of nodes $v_1, v_2 \in V^*$, there exists a finite directed path which starts in v_1 and ends in v_2. A strongly connected component is the subgraph induced by a strongly connected set that is maximal. This means the component cannot be extended without ceasing to be strongly connected. The formal definitions of these properties are as follows:

Definition 1.8: A set of nodes $V^* \subseteq V$ is **strongly connected** iff:

(i) $\forall (v_1,v_2) \in V^* \times V^*$: $DPF(v_1,v_2) \neq \emptyset$.

A **strongly connected component** is the subgraph induced by a non-empty set of nodes $V^* \subseteq V$, where:

(ii) V^* is strongly connected.
(iii) $\forall V' \subseteq V$: (V' strongly connected $\wedge V^* \subseteq V') \Rightarrow V^* = V'$.

A strongly connected component is **trivial** iff it has a single node and no arcs. The set of all strongly connected components is denoted by SCC.

Proposition 1.9: The set of nodes of the strongly connected components SCC is a partition of the set of all nodes V.

Proof: The proof is rather straightforward. It can be found in most textbooks on graph theory. Hence it is omitted. □

For a node $v \in V$ and a component $c \in SCC$ we use the notation $v \in c$ to denote that v is one of the nodes in c. A similar notation is used for arcs.

From Prop. 1.9 it follows that each node is contained in exactly one of the strongly connected components. We use v^c to denote the component to which $v \in V$ belongs, and we also use $X^c = \{v^c \in SCC \mid v \in X\}$ where $X \subseteq V$.

Finally, we use $DCS(c) \subseteq DCS$ to denote the set of all simple directed cycles where all nodes and arcs belong to the component $c \in SCC$.

Now we are ready to define the SCC-graph of a directed graph DG. The SCC-graph contains a node for each strongly connected component (of DG). The SCC-graph contains those arcs (among the arcs of DG) which connect two different strongly connected components. Intuitively, we can obtain the SCC-graph by folding the original graph. We position all nodes of each strongly connected component "on top of each other" and we lump them into a single node that has all the input and output arcs of the original nodes – with the exception of those arcs which start and end in the same component.

Definition 1.10: The directed graph $DG^* = (V^*, A^*, N^*)$ is the **SCC-graph** of DG iff the following properties are satisfied:

(i) $V^* = SCC$.
(ii) $A^* = \{a \in A \mid s(a)^c \neq d(a)^c\}$.
(iii) $\forall a \in A^*$: $N^*(a) = (s(a)^c, d(a)^c)$.

Definitions 1.8 and 1.10 are illustrated by Fig. 1.3. The thick nodes and arcs constitute the SCC-graph of the directed graph containing the thin nodes and arcs (for readability we have omitted all inscriptions). It should be noted that the SCC-graph may have several arcs between two components (as illustrated, e.g., by the two upper components).

For CP-nets with a cyclic behaviour, we often have O-graphs with very small SCC-graphs. Most CP-nets considered in Vol. 1 have an O-graph with a single strongly connected component, and hence the SCC-graph has a single node and no arcs. For acyclic CP-nets, we do not gain anything by considering the strongly connected components. For an acyclic directed graph, all strongly connected components are trivial and this implies that the SCC-graph is isomorphic to the original graph.

The following proposition formulates a number of standard properties of SCC-graphs. The properties will be used in the proofs of the proof rules (see Sects. 1.4, 2.3, and 3.2). We use SCC_T to denote the set of all **terminal** strongly connected components (i.e., all those components with no output arcs).

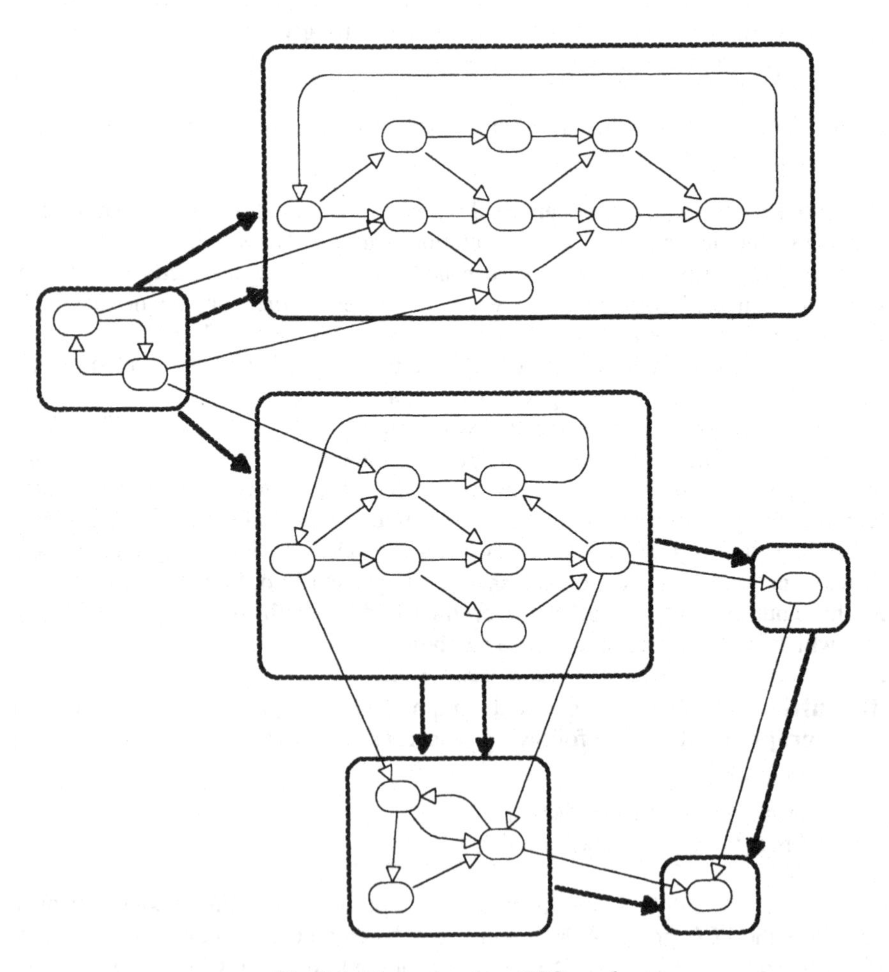

Fig. 1.3. Example of an SCC-graph

Proposition 1.11: The following properties are satisfied:

(i) The SCC-graph is acyclic.
(ii) V finite $\Rightarrow$ SCC finite.
(iii) V finite $\Rightarrow$ $\forall c_1 \in SCC\ \exists c_2 \in SCC_T$: $DPF(c_1,c_2) \neq \emptyset$.
(iv) $\forall v_1,v_2 \in V$: $DPF(v_1,v_2) \neq \emptyset \Leftrightarrow DPF(v_1{}^c,v_2{}^c) \neq \emptyset$.

Proof: **Property (i):** Assume that the SCC-graph has a cycle:

$$c_1\ a_1\ c_2\ a_2\ c_3 \ldots c_n\ a_n\ c_{n+1}.$$

A cycle must contain at least one arc and since SCC-graphs do not have arcs where the source and destination are identical, we conclude that the cycle contains at least two different components. By means of the cycle, it is easy to prove that the nodes of the components $\{c_1, c_2, \ldots, c_{n+1}\}$ together form a strongly connected set. However, this is in conflict with the assumption that the nodes of each c_i is a *maximal* strongly connected set.

Property (ii): The strongly connected components are non-empty and pairwise disjoint. This means that $|SCC| \leq |V|$.

Property (iii): We shall construct a finite directed path that starts with c_1 and ends with a terminal component. Iff $c_1 \in SCC_T$ we use the path c_1 (of length zero). Otherwise, c_1 has an output arc a_1 with destination $c_2 \in SCC$. Iff $c_2 \in SCC_T$ we use the path $c_1\ a_1\ c_2$. Otherwise, c_2 has an output arc a_2 with destination $c_3 \in SCC$. Iff $c_3 \in SCC_T$ we use the path $c_1\ a_1\ c_2\ a_2\ c_3$. And so on. The argument continues until we reach a component in SCC_T. This must happen after a finite number of steps, due to (i) and (ii).

Property (iv): This property follows from the definition of the SCC-graph and the definition of strongly connected sets. □

1.4 Proof Rules for O-graphs

In this section we assume that we are dealing with a CP-net which has a *finite* O-graph OG = (V, A, N). When this is the case, we can use the O-graph to investigate the dynamic properties introduced in Chap. 4 of Vol. 1. This is done by means of a number of proof rules. Each proof rule states a relationship between a dynamic property of the CP-net and a property of the corresponding O-graph. For CP-nets with a cyclic behaviour, it is often much more efficient to investigate the corresponding SCC-graph, and thus we shall also consider proof rules that state a relationship between a dynamic property of the CP-net and a property of the SCC-graph. Some of the proofs in this section are a bit complicated and may be skipped by readers who are primarily interested in the practical application of CP-nets.

First we consider the reachability properties, i.e., those directly based on the enabling and occurrence rules. The reachability properties are defined in Sects. 2.3 and 3.4 of Vol. 1.

Proposition 1.12: For the **reachability properties** we have the following proof rules, valid for all $M_1,M_2 \in [M_0\rangle$:

(i) $[M_0\rangle = V$.
(ii) $M_2 \in [M_1\rangle \Leftrightarrow DPF(M_1,M_2) \neq \emptyset$.
(iii) $M_2 \in [M_1\rangle \Leftrightarrow DPF(M_1{}^c,M_2{}^c) \neq \emptyset$.
(iv) $M_2 \in [M_1\rangle \Leftarrow |SCC| = 1$.

Explanation: Property (i) tells us that we can determine whether a node is reachable by investigating whether it appears as a node. Property (ii) tells us that a marking M_2 is reachable from another marking M_1 iff the O-graph has a directed path from M_1 to M_2. Property (iii) tells us that, instead of looking for directed paths in the O-graph, we can look for directed paths in the corresponding SCC-graph, which is often much smaller. A marking M_2 is reachable from another marking M_1 iff the SCC-graph has a directed path from the component to which M_1 belongs to the component to which M_2 belongs. Finally, property (iv) investigates the situation where the O-graph has only one strongly connected component. We then know that all markings that are reachable are also reachable from each other.

It is important to notice that properties (ii)–(iv) only give us information about markings that are reachable from the initial marking. If we want to get information about non-reachable markings, we must choose another initial marking and construct a new O-graph.

Proof: Property (i) is an immediate consequence of Def. 1.3 (i). Property (ii) follows from Prop. 1.6. Property (iii) follows from (ii) and from Prop. 1.11 (iv). Finally, property (iv) follows from (ii) and from Def. 1.8 (i). □

Next, let us consider the boundedness properties defined in Sect. 4.2 of Vol. 1. For $A \subseteq S_{MS}$ we define $\max(A) \in S_{MS}$ to be the multi-set in which:

$$(\max(A))(s) = \max_{m \in A} m(s)$$

for all $s \in S$. We define min(A) analogously. In general, we do not know that max(A) and min(A) belong to A itself. In mathematical terms, this means max(A) is the supremum of A while min(A) is the infimum.

Proposition 1.13: For the **boundedness properties** we have the following proof rules, valid for all $X \subseteq TE$, all $p \in PI$, and all functions $F \in [\mathbb{M} \to A]$ where $(A,\leq)$ is an arbitrary set with a linear ordering relation:

(i) $\text{BestUpperBound}(X) = \max_{M \in V} |(M|X)|$.
(ii) $\text{BestUpperIntegerBound}(p) = \max_{M \in V} |M(p)|$.
(iii) $\text{BestUpperMulti-setBound}(p) = \max_{M \in V} M(p)$.
(iv) $\text{BestUpperBound}(F) = \max_{M \in V} F(M)$.

By replacing max with min, we get four rules for lower bounds.

Explanation: In this section, we only consider CP-nets with a finite O-graph. This means the CP-net has a finite set of reachable markings and hence we know that there exist upper bounds for all different kinds of boundedness properties. The proof rules show that we can use the O-graph to obtain the best possible upper and lower bounds.

In Sect. 4.2 of Vol. 1 we also defined bounds for places. However, these bounds are derived from the bounds of place instances, and thus they are covered by Prop. 1.13 (ii)–(iii). A similar remark applies to the corresponding proof rules for OE-graphs and OS-graphs (in Sects. 2.3 and 3.2).

Proof: The proof follows from Prop. 1.12 (i) and from the various boundedness definitions in Sect. 4.2 of Vol. 1. The ordering relation on A must be linear in order to guarantee that the maximum and minimum exists. □

Next, let us consider the home properties defined in Sect. 4.3 of Vol. 1. We use HS to denote the set of all home spaces, while we use HM to denote the set of all home markings.

Proposition 1.14: For the **home properties** we have the following proof rules, valid for all $X \subseteq [M_0\rangle$ and all $M \in [M_0\rangle$:

(i) $X \in HS \Leftrightarrow SCC_T \subseteq X^c$.
(ii) $X \in HS \Rightarrow |SCC_T| \leq |X|$.

(iii) $M \in HM \Leftrightarrow SCC_T = \{M^c\}$.
(iv) $HM \neq \emptyset \Leftrightarrow |SCC_T| = 1$.
(v) $M_0 \in HM \Leftrightarrow |SCC| = 1$.

Explanation: Property (i) tells us that a set of markings is a home space iff it contains a node from each of the terminal strongly connected components. This means a home space must contain at least as many markings as there are terminal strongly connected components. This is stated in property (ii).

Properties (iii) and (iv) are similar to properties (i) and (ii). They deal with home markings, i.e., home spaces which have only a single marking. There exist home markings iff the O-graph has only a single terminal component – in which case each marking of the terminal component is a home marking. Finally, property (v) deals with the particular case in which there is only one strongly connected component.

Proof: **Property (i):** Assume that $SCC_T \subseteq X^c$ and let $M' \in [M_0\rangle$ be a reachable marking. We shall then prove that $X \cap [M'\rangle \neq \emptyset$ (see Def. 4.8 (ii) of Vol. 1).

From Prop. 1.11 (iii) we know there exists a terminal component $c \in SCC_T$ such that $DPF(M'^c, c) \neq \emptyset$. From our assumption we know that c contains a marking $M \in X$. Thus we have $DPF(M'^c, M^c) \neq \emptyset$ and by Prop. 1.12 (iii) we conclude that $M \in [M'\rangle$. Hence $M \in X \cap [M'\rangle$.

Next assume that there exists a terminal component $c \in SCC_T$ with no elements from X. Let $M \in c$ be a node in this component. It is then easy to see that M is a reachable marking from which it is impossible to reach a marking in X (because it is impossible to find a directed path which leaves the terminal component c).

Property (ii): Assume that $X \in HS$. We then have:

$$|SCC_T| \leq |X^c| \leq |X|$$

where the first inequality follows from $SCC_T \subseteq X^c$ and the second follows from the definition of X^c.

Properties (iii)–(v): These properties follow from (i) and (ii) and the fact that the SCC-graph of a finite graph has at least one terminal component. The latter follows from Prop. 1.11 (iii). □

We now introduce some notation that allows us to inspect how the different binding elements appear in O-graphs and the corresponding SCC-graphs. For a node $M_1 \in V$ we use $BE(M_1)$ to denote the set of all binding elements which appear in an output arc of M_1:

$$BE(M_1) = \{b \in BE \mid \exists M_2 \in V: (M_1,b,M_2) \in A\}.$$

For a directed path $dp \in DP$ we use $BE(dp)$ to denote the set of all binding elements which appear in an arc of dp:

$$BE(dp) = \{b \in BE \mid \exists M_1,M_2 \in dp: (M_1,b,M_2) \in dp\}.$$

For a strongly connected component $c \in SCC$ we use $BE(c)$ to denote the set of all binding elements which appear in an arc which starts in c:

$$BE(c) = \{b \in BE \mid \exists M_1 \in c \; \exists M_2 \in V: (M_1,b,M_2) \in A\}.$$

Next, let us consider the liveness properties defined in Sects. 4.4 of Vol. 1.

Proposition 1.15: For the **liveness properties** we have the following proof rules, valid for all $M \in [M_0\rangle$, all $X \subseteq BE$, and all $t \in T \cup TI$:

(i) M is dead $\Leftrightarrow$ M is terminal.
(ii) M is dead $\Leftrightarrow$ M^c is terminal and trivial.

(iii) X is dead in M $\Leftrightarrow$ $\forall c \in SCC$: $[DPF(M^c,c) = \emptyset \vee BE(c) \cap X = \emptyset]$.
(iv) X is live $\Leftrightarrow$ $\forall c \in SCC_T$: $BE(c) \cap X \neq \emptyset$.

(v) t is live $\Leftrightarrow$ $\forall c \in SCC_T$: $t \in BE(c)$.
(vi) t is strictly live $\Leftrightarrow$ $\forall c \in SCC_T \; \forall b \in B(t)$: $(t,b) \in BE(c)$.

Explanation: Property (i) tells us that a reachable marking is dead iff the corresponding node of the O-graph is terminal, i.e., has no output arcs. Property (ii) tells us that, instead of looking for terminal nodes of the O-graph, we can look for terminal and trivial nodes of the corresponding SCC-graph.

Property (iii) tells us that a set of binding elements is dead in a marking M iff none of the strongly connected components, which can be reached from the component of M, contain a node which has a binding element from X in an output arc. Finally, property (iv) tells us that a set of binding elements is live iff each terminal strongly connected component contains a node which has a binding element from X in an output arc.

Properties (v) and (vi) are particular cases of (iv). It should be noted that they are satisfied both for transitions and for transition instances. We use $t \in BE(c)$ to denote that the t appears in a binding element b which belong to BE(c).

In Sect. 4.4 of Vol. 1 we also defined strict liveness and liveness with respect to an equivalence relation. These definitions are derived from the basic liveness definition, and thus they are covered by Prop. 1.15 (iv). A similar remark applies to the corresponding proof rules for OE-graphs and OS-graphs.

Proof: **Property** (i): Straightforward consequence of Prop. 1.12 (ii).

Property (ii): It is easy to see that a node $M \in V$ is terminal iff it belongs to a strongly connected component that is terminal and has only one node and no arcs. This means the right-hand sides of (i) and (ii) are equivalent.

Property (iii): Assume that there exists a strongly connected component $c \in SCC$ that does not fulfil the requirement in the right-hand side of (iii). From $BE(c) \cap X \neq \emptyset$ we know there exists a marking $M' \in c$ which has a binding element $b \in X$ in an output arc. According to Def. 1.3 (ii) this means b is enabled in M'. From $DPF(M^c,c) \neq \emptyset$ and Prop. 1.12 (iii) we know that $M' \in [M\rangle$. Hence we have shown that X is non-dead in M.

Next assume that X is non-dead in M. This means there exists a finite occurrence sequence which starts in M and ends in a marking M' in which a binding element from X is enabled. From Prop. 1.12 (iii) we know that $DPF(M^c,M'^c) \neq \emptyset$ and from Def. 1.3 (ii) we know that $BE(M'^c) \cap X \neq \emptyset$. Hence we have shown that M'^c does not satisfy the right-hand side of (iii).

Property (iv): Assume that $BE(c) \cap X \neq \emptyset$ for all terminal strongly connected components and let M be a reachable marking. We shall prove that X is non-dead in M (see Def. 4.10 (iii) of Vol. 1). From Prop. 1.11 (iii) we conclude that there exists a terminal strongly connected component c such that $DPF(M^c,c) \neq \emptyset$. This means the following is false:

$$DPF(M^c,c) = \emptyset \vee BE(c) \cap X = \emptyset$$

and thus we conclude from (iii) that X cannot be dead in M.

Next assume that there exists a terminal strongly connected component c such that $BE(c) \cap X = \emptyset$. From (iii) it is easy to prove that X is dead in all $M \in c$. Hence we conclude that X cannot be live.

Properties (v)–(vi): Direct consequences of property (iv) and the definitions of liveness and strict liveness for a transition/transition instance. □

Finally, let us consider the fairness properties defined in Sects. 4.5 of Vol. 1. For a set of binding elements $X \subseteq BE$, we construct a subgraph of the O-graph by deleting all arcs that contain a binding element from X. We use $SCC_{O\setminus X}$ to denote the set of all strongly connected components of this subgraph. Moreover, we use the notation defined prior to Prop. 1.15.

Proposition 1.16: For the **fairness properties** we have the following proof rules, valid for all $X \subseteq BE$:

(i) X is impartial $\Leftrightarrow$ $\forall dc \in DCS$: $[BE(dc) \cap X \neq \emptyset]$.
(ii) X is fair $\Leftrightarrow$ $\forall dc \in DCS$: $[BE(dc) \cap X \neq \emptyset \vee \forall M \in dc\colon BE(M) \cap X = \emptyset]$.
(iii) X is just $\Leftrightarrow$ $\forall dc \in DCS$: $[BE(dc) \cap X \neq \emptyset \vee \exists M \in dc\colon BE(M) \cap X = \emptyset]$.

(iv) X is impartial $\Leftrightarrow$ $\forall c \in SCC_{O \setminus X}$: [c is trivial].
(v) X is fair $\Leftrightarrow$ $\forall c \in SCC_{O \setminus X}$: $[c \text{ is trivial} \vee \forall M \in c\colon BE(M) \cap X = \emptyset]$.
(vi) X is just $\Leftrightarrow$ $\forall c \in SCC_{O \setminus X}$: [c is trivial $\vee$
$\forall dc \in DCS(c)\ \exists M \in dc\colon BE(M) \cap X = \emptyset]$.

Explanation: Property (i) tells us that a set of binding elements X is impartial iff all simple directed cycles contain one or more arcs with a binding element from X. Property (ii) tells us that X is fair iff all simple directed cycles either contain one or more arcs with a binding element from X or contain only nodes where no element from X appears in the output arcs. Property (iii) tells us that X is just iff all simple directed cycles either contain one or more arcs with a binding element from X or contain at least one node where no element from X appears in the output arcs.

It should be noted that property (ii) and (iii) are nearly identical. This is rather surprising, because the definitions of just and fair seem to be quite different from each other; see Sect. 4.5 of Vol. 1.

Properties (iv)–(vi) can easily be deduced from properties (i)–(iii), respectively. Property (iv) tells us that X is impartial iff all nodes in $SCC_{O \setminus X}$ are trivial components, i.e., have a single node and no arcs. Property (v) tells us that X is fair iff all components in $SCC_{O \setminus X}$ are either trivial or only contain nodes where no element from X appears in the output arcs. Finally, property (vi) tells us that X is just iff all components in $SCC_{O \setminus X}$ are either trivial or contain no simple directed cycles where all nodes have a binding element from X in an output arc.

Properties (v) could have been formulated in a way that looks more similar to property (vi). Then the right-hand side of (v) would have been identical to the right-hand side of (vi), except that $\exists M \in dc$ is replaced by $\forall M \in dc$. We have chosen a simpler formulation for (v). From the properties of strongly connected components is easy to see that the two formulations are equivalent.

In Sect. 4.5 of Vol. 1 we also defined fairness properties for transitions and transition instances. Moreover, we defined strict fairness properties and fairness properties with respect to an equivalence relation. All these definitions are derived from the basic fairness properties, and thus they are covered by Prop. 1.16. A similar remark applies to the corresponding proof rules for OE-graphs and OS-graphs.

Finally, it should be noted that the second line of (vi) can be tested by considering the SCC-graph of the subgraph of c induced by those M for which $BE(M) \cap X \neq \emptyset$.

Proof: **Property** (i): Assume that each simple directed cycle contains an element from X, and let $\sigma \in OSI$ be an infinite occurrence sequence (which starts in a reachable marking). We shall prove that $OC_X(\sigma) = \infty$ (see Def. 4.12 (i) of

Vol. 1). From Prop. 1.6 (i), we know that σ has a matching infinite directed path $dp \in DPI$. Since dp is infinite it must contain one of the directed cycles $dc \in DCS$ infinitely many times. From $BE(dc) \cap X \neq \emptyset$ we then conclude that $OC_X(\sigma) = \infty$.

Next, assume that X is impartial and let $dc \in DCS$ be a simple directed cycle. We shall prove that dc contains an element from X. To do this we construct an infinite occurrence sequence $\sigma \in OSI$ that matches the cycle dc repeated infinitely often. This can be done due to Prop. 1.6 (ii). From the definition of impartiality, we know that $OC_X(\sigma) = \infty$. Hence we conclude that $BE(dc) \cap X \neq \emptyset$.

Properties (ii)–(iii): The proofs of (ii) and (iii) are rather straightforward modifications of the proof of (i). Hence they are omitted.

Properties (iv)–(vi): The O-graph contains a simple directed cycle without elements from X iff $SCC_{O \setminus X}$ has a non-trivial component. With this observation it is easy to verify that (iv)–(vi) follow from (i)–(iii), respectively. □

1.5 Example of O-graphs: Distributed Data Base

In this section we study the O-graph for the data base system from Sect. 1.3 of Vol. 1. For this system it is rather easy to calculate the number of nodes and arcs in the O-graph, and thus we can study how the size of the O-graph grows when we increase the number of data base managers.

The O-graph for three data base managers $DBM = \{d_1, d_2, d_3\}$ is shown in Fig. 1.4. Each node represents a reachable marking. To save space (in our drawing) we represent each marking by listing those data base managers that have a message addressed to them – on *Sent*, *Received*, or *Acknowledged*, respectively. This means (2,3,–) denotes a marking in which d_2 is *Inactive* and d_3 *Performing* while d_1 is *Waiting*. Analogously (23,–,–) denotes a marking in which d_2 and d_3 are *Inactive*, while d_1 is *Waiting*. The initial marking is represented by (–,–,–). This node is drawn twice to avoid long arcs. The second copy has a dashed border line. We invite the reader to check that the information in these triples is sufficient to calculate the entire marking.

Each arc represents a triple (M_1,b,M_2) where M_1 and M_2 are reachable markings and b is a binding element. As explained immediately below Def. 1.3, there is no need to include M_1 and M_2 in the arc inscription. The binding element is represented as a pair where the first element is the transition while the second is the binding. The transition names are abbreviated to SM, RM, SA, and RA. We write SM, i and RM, i, k instead of $(SM, <s{=}d_i>)$ and $(RM, <s{=}d_i, r{=}d_k>)$, and analogously for SA and RA.

From the O-graph in Fig. 1.4 it is possible to verify all the dynamic properties postulated for the data base system in Chap. 4 of Vol. 1 (see Exercise 1.2).

Now let us calculate the size of the O-graph as a function of the number of data base managers n (we assume that $n \geq 3$). First we calculate the number of nodes. To do this we observe that there exists a reachable marking for each way we can select one data base manager to be *Waiting* and partition the remaining $n-1$ data base managers into the three sets we use to represent the marking.

From the theory of probability it is known that there are $n*3^{n-1}$ such markings. This can be seen as follows. First we choose the *Waiting* data base manager (n possibilities). Then there are three choices for each of the remaining $n-1$ managers (3^{n-1} possibilities). The only other reachable marking that is possible is the marking in which all data base managers are inactive, i.e., the initial marking. Hence we conclude that the O-graph has the following number of nodes:

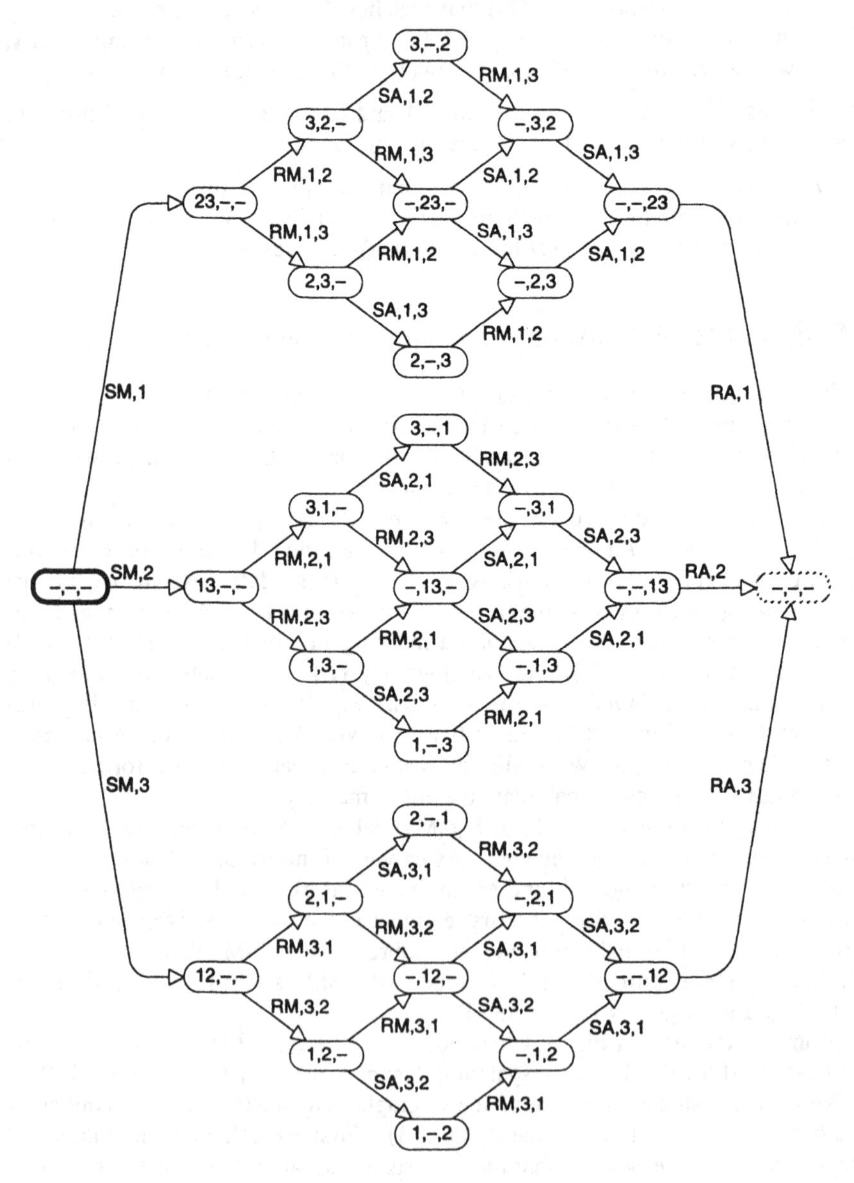

Fig. 1.4. O-graph for data base system with 3 managers

$$N_O(n) = 1 + n*3^{n-1} = O(n*3^n)$$

where the notation $O(n*3^n)$ means that $N_O(n)$ grows with the same speed as $n*3^n$ when n increases. More formally, we say that the complexity of $N_O(n)$ is of **order** $n*3^n$. A formal definition of order can be found in most text books on complexity theory.

Next we calculate the number of arcs. Each reachable marking $M_1 = (S,R,A)$, has an enabled binding of RM for each element in S and has an enabled binding of SA for each element of R. This gives us $|S|+|R|$ different output arcs of M_1. Next we inspect the markings $M_2 = (R,A,S)$ and $M_3 = (A,S,R)$ that we obtain from M_1 by cyclic rotations of the three sets S, R, and A. It is easy to see that M_2 and M_3 are also reachable. Together M_1, M_2, and M_3 have:

$$|S|+|R| + |R|+|A| + |A|+|S| = 2*(|S|+|R|+|A|) = 2*(n-1)$$

output arcs containing RM or SA.

Let us for a moment ignore the initial marking. Then we can partition all the other reachable markings in such a way that each component of the partition contains three markings which have a similar relationship to each other as M_1, M_2, and M_3. There are $(N_O(n)-1)/3$ such components and each of these has $2*(n-1)$ output arcs. Together they have the following number of output arcs:

$$2*(n-1)*(N_O(n)-1)/3 = 2*(n-1)*n*3^{n-1}/3 = 2*(n-1)*n*3^{n-2}.$$

The only other arcs in the O-graph are the $2*n$ arcs which surround M_0, and thus we get the following number of arcs:

$$A_O(n) = 2*n + 2*(n-1)*n*3^{n-2} = O(n^2*3^n).$$

The results for different numbers of data base managers are shown in Fig. 1.5. They illustrate the **space complexity** of the O-graph algorithm in Prop. 1.4.

As illustrated above, it is often the case that the O-graph of a CP-net grows very fast when the sizes of the involved colour sets increase. However, in practice, it is fortunately often sufficient to consider rather small colour sets in order

$\|DBM\|$	$N_O(n)$	$A_O(n)$
$O(n)$	$O(n*3^n)$	$O(n^2*3^n)$
2	7	8
3	28	42
4	109	224
5	406	1,090
6	1,459	4,872
7	5,104	20,426
8	17,497	81,664
9	59,050	314,946
10	196,831	1,181,000
15	71,744,536	669,615,690
20	23,245,229,341	294,439,571,680

Fig. 1.5. The size of the O-graphs for the data base system

to verify the logical correctness of a given CP-net. Having convinced ourselves that the data base system or the telephone system has the correct behaviour for 4 or 5 managers/phones, we can feel pretty sure that the CP-nets are also correct for any larger number of managers/phones. Sadly, a similar statement is not true when we try to evaluate the performance of a given system.

1.6 Example of O-graphs: Dining Philosophers

In this section we study the O-graph for the philosopher system from Exercise 1.6 of Vol. 1. Five Chinese philosophers are sitting around a circular table. In the middle of the table there is a delicious dish of rice, and between each pair of philosophers there is a single chopstick. Each philosopher alternates between thinking and eating. To eat, the philosopher needs two chopsticks, and he is only allowed to use the two which are situated next to him (on his left and right side). The sharing of chopsticks prevents two neighbours from eating at the same time.

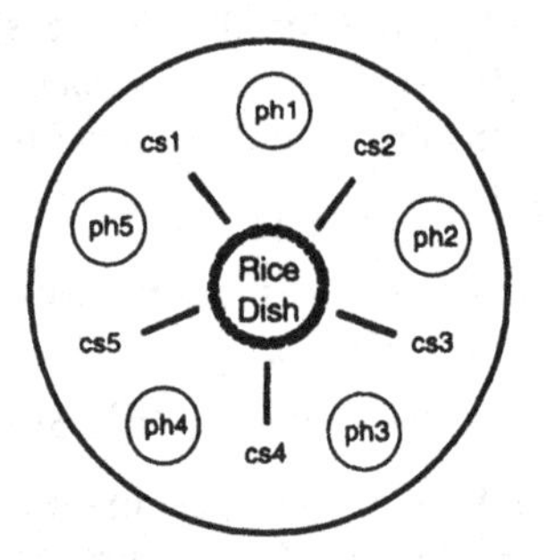

The philosopher system is modelled by the CP-net in Fig. 1.6. The PH colour set represents the philosophers, while the CS colour set represents the chopsticks. The function *Chopsticks* maps each philosopher into the two chopsticks next to him.

The O-graph for a system with five philosophers is shown in Fig. 1.7. Each node represents a reachable marking. To improve the overview, we have indicated the marking by means of a small picture which shows the eating philosophers. As an example, the uppermost node represents the marking in which ph_2 and ph_5 are eating, while the remaining philosophers are thinking. Each arc is labelled by a philosopher ph_i and there are always two arcs on top of each other – one in each direction. One of the arcs has (TakeChopsticks, <p=ph(i)>) as binding element, while the other arc has (PutDownChopsticks, <p=ph(i)>) as binding element. The direction of an arc follows from the binding element. An arc with *Take Chopsticks* increases the number of eating philosophers, while an arc with *Put Down Chopsticks* decreases the number.

For the philosopher system it can be shown that the number of nodes grows as a Fibonacci sequence, i.e., that $N_O(n) = N_O(n-1)+N_O(n-2)$ where $N_O(2) = 3$ while $N_O(3) = 4$. It can also be shown that the number of arcs is $A_O(n) = 2*n*F(n)$ where $F(n)$ is the Fibonacci sequence determined by

F(2) = F(3) = 1. The results are shown for different numbers of philosophers in Fig. 1.8. Analogously to the data base system, it can be seen that the O-graph grows very fast when the sizes of the involved colour sets increase.

```
val n = 5;
color PH = index ph with 1..n   declare ms;
color CS = index cs with 1..n   declare ms;
var p : PH;
fun Chopsticks(ph(i)) = 1`cs(i)+1`cs(if i=n then 1 else i+1);
```

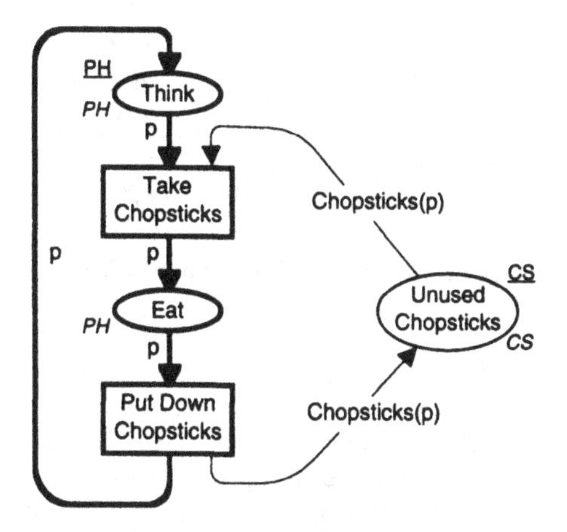

Fig. 1.6. CP-net describing the philosopher system

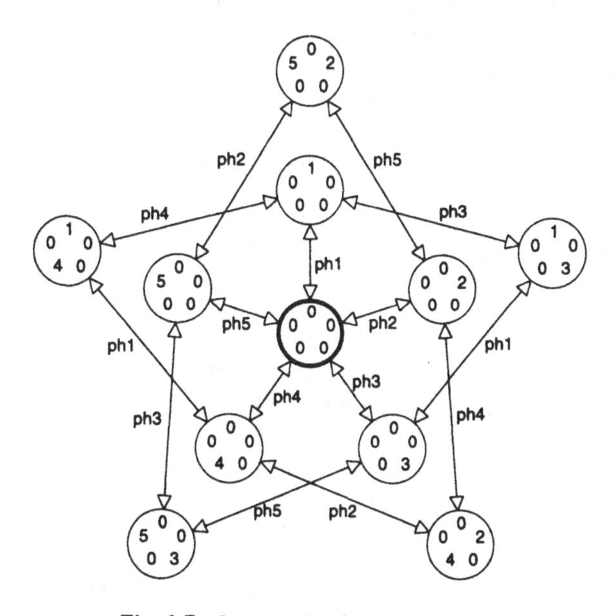

Fig. 1.7. O-graph for five philosophers

In Chaps. 2 and 3, we shall describe methods that allow us to construct much smaller occurrence graphs – without losing the possibility of investigating the dynamic properties of the CP-net.

|PH|	$N_O(n)$	$A_O(n)$
2	3	4
3	4	6
4	7	16
5	11	30
6	18	60
7	29	112
8	47	208
9	76	378
10	123	680
15	1,364	11,310

Fig. 1.8. The size of the O-graphs for the philosopher system

1.7 Computer Tools for O-graphs

This section describes the basic ideas behind a tool that supports the construction and analysis of O-graphs. The occurrence graph tool is integrated with the CPN editor and CPN simulator described in Chap. 6 of Vol. 1. The occurrence graph tool is rather new, and thus we have less experience with it than we have with the CPN editor and simulator. Although the section describes a concrete tool, it can also be seen as a presentation of a set of design criteria and design ideas that are relevant to all occurrence graph tools.

The occurrence graph tool can construct the O-graph of an arbitrary hierarchical CP-net, provided that the O-graph has a reasonable size. With the present version of the occurrence graph tool we have constructed O-graphs that have up to 100,000 nodes and 500,000 arcs. With later versions, we expect to be able to handle much larger graphs. The data base system with eight data base managers has an O-graph with 17,500 nodes and 81,600 arcs; see Fig. 1.5. The construction of this O-graph takes approximately 60 minutes on a Macintosh IIfx or a SUN 4, or 30 minutes on a DEC Station 5000/240. The limitation is not only determined by the time complexity, but also by the space complexity. The above O-graph for eight data base managers uses 60 MB of memory. The occurrence graph tool is implemented in Standard ML (SML) and the constructed O-graph is a complex SML data structure.

At a first glance, the above figures may seem rather discouraging. However, it should be noted that the analysis of an O-graph is usually much faster than the construction. This is in particular true when we deal with a cyclic CP-net for which, with great benefit, we can construct the SCC-graph. Even a large O-graph can be constructed and analysed in less than one day, totally automatically. The alternative may be a lengthy and error-prone manual testing and debugging, e.g., by means of a set of simulations. Such an approach is often much more expensive than the occurrence graph analysis – and less reliable.

Some years ago a company used several weeks of computing power to construct an O-graph for a PT-net, modelling a protocol to be used in a new personal computer. According to the company, the cost of the O-graph construction and analysis was insignificant compared to the potential loss caused if a faulty piece of software had to be replaced in thousands of computers distributed all over the world.

Integration with the CPN simulator

The construction of the O-graph uses exactly the same input as the CPN simulator. It is possible to make an O-graph for part of a large model. This is done in the same way as in the CPN simulator, i.e., by means of the prime pages and one of the mode attributes (see Sect. 6.2 of Vol. 1). An O-graph can be constructed with or without time and with or without code segments. However, code segments should be used with care. If they have side effects that must be executed in a particular order, it makes no sense to construct an O-graph (because this may execute the code segments in the wrong order).

The present version of the occurrence graph tool is a separate program. However, the next versions will be more tightly integrated with the CPN simulator. This means the simulator will be able to perform an automatic simulation of an occurrence sequence found using the occurrence graph tool. It also means that the occurrence graph tool will be able to refer to the current marking of the simulator (and, e.g., search for nodes which have an identical or similar marking).

Construction of O-graphs

To construct an O-graph it is necessary to be able to process a marking, i.e., to find the set of all enabled binding elements and the corresponding direct successor markings. However, this is not a problem, because it is done in exactly the same way as in the CPN simulator. The only difference is that we cannot allow the set of enabled binding elements to be infinite (or so big that we run into complexity problems).

In the CPN simulator we only have one marking at a time. In the occurrence graph tool we may have several thousands markings in the same O-graph. Moreover, a lot of these markings will be almost identical. Hence we may save a significant amount of memory if we avoid duplication of identical parts. To do this we represent each marking as a set of pointers, as shown in Fig. 1.9. This means each multi-set only appears once – even though it may appear in many different markings and for many different places.

When we construct a new marking from an existing marking, we only change the marking of a few pages. This means we can reuse most of the page records. In Fig. 1.9, the marking *MarkK* shares the page records *Page1* and *PageN* with *Mark1*, while a new page record, *Page2a*, has been created for the second page. The new page record shares some multi-sets with *Page2* and some multi-sets with other page records. Hence it has only been necessary to create one new multi-set record (the second from the bottom).

At first glance the reader may think that the degree of sharing (reuse of records) in Fig. 1.9 is atypical. However, this is not the case. As an example, *MarkK* could be a marking obtained from *Mark1* by the occurrence of a transition which belongs to *Page2*. If the transition does not have any input/output places which are ports, sockets, or members of fusion sets, it is obvious that only the record for *Page2* needs to be changed, while all the other page records can be reused. A similar observation can be made for the multi-set records. Here, only the records corresponding to input/output places of the occurring transition need to be changed.

A data base system with 5 managers has 406 reachable markings (see Fig. 1.5). Each of these markings consists of 9 multi-sets (one for each place). This means the markings together contain 3654 multi-sets. However, only 116 of these are different from each other. There are 2 possible multi-sets for E and $2^5 = 32$ multi-sets for DBM. Finally, there are 6 possible MES multi-sets at *Unused* while there are $(5 * 2^4) - 4 = 76$ possible MES multi-sets at *Sent*, *Received*, and *Acknowledged*. The latter can be seen by noticing that each of the 5 managers has 4 messages to send. This gives us $5 * 2^4$ subsets of messages – of which 5 are identical to Ø. This means we only need 3.17 % of the space we would need without the multi-set sharing (in this calculation we ignore the space used for the pointers and we assume that all multi-sets use the same amount of space). For a system with 10 data base managers the savings are even more dramatic. Here, we have nearly 200,000 markings, i.e., 1.8 million multi-sets.

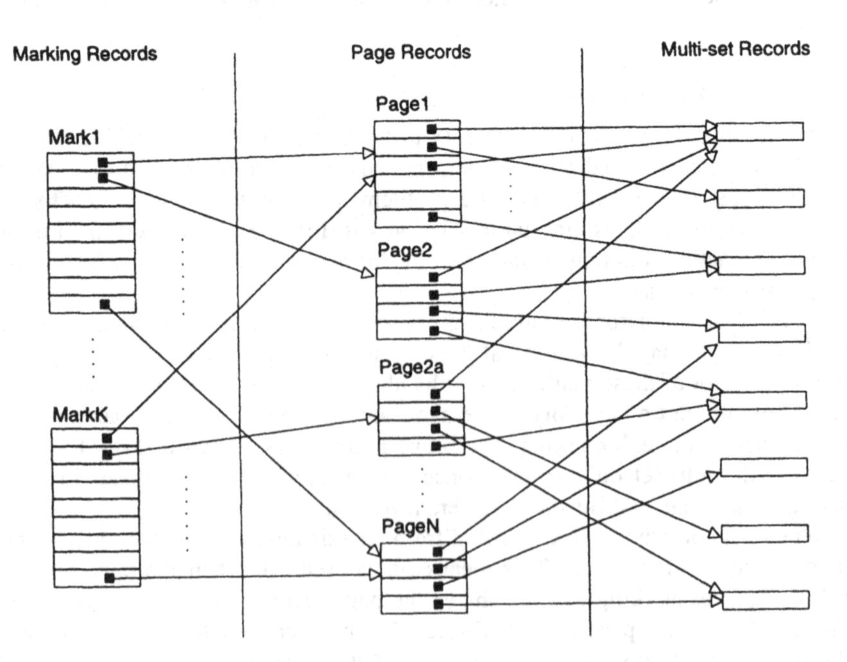

Fig. 1.9. Representation of a set of markings

However, only 6,148 are different from each other, and this means we only need 0.35 % of the space we would need without the multi-set sharing.

The above strategy saves a lot of space. However, it also makes the construction of the O-graph much more efficient, because it allows a more efficient test for identity of markings. Whenever a new marking has been constructed, we must check whether the marking is identical to any of the existing markings. To do this we keep the markings in a search tree. Moreover, we have a search tree for each kind of multi-set, i.e., for each colour set. We also have a search tree for each kind of page record, i.e., for each page. To be positioned in a search tree, each record is given an integer key calculated from the contents of the corresponding marking, page record, or multi-set. Multi-set records have a key that is a function of the number of tokens in the multi-set, or the coefficients which appear in the multi-set. Page records have a key that is a function of the keys of the corresponding multi-set records. Analogously, marking records have a key that is a function of the keys of the corresponding page records.

When a binding element occurs, some of the surrounding places get a modified multi-set.

For each modified multi-set we investigate the corresponding search tree and insert the multi-set, unless it is already present. The use of the integer keys means we only have to compare the multi-set to those multi-sets that have an identical key. This increases the efficiency of the algorithm.

For each modified page record we investigate the corresponding search tree. When the page record contains a pointer to a new multi-set (one which was not present in its search tree) we know that the multi-set record cannot be identical to any of the existing page records. Hence we can insert the page record, without comparing it with any existing page records. Otherwise, we must (before insertion) compare the page record to those records which have the same key. This is done by comparing the pointers in the two page records (which is a fast operation).

For a marking record the situation is analogous. When a new page record or a new multi-set record has been created, we know that the marking must be new. Hence we can insert the marking record without any comparison with existing marking records. Otherwise, we have to compare with those marking records that have the same key. This is done by comparing the pointers in the two marking records.

Binding elements are also saved in a search tree. For O-graphs this is primarily done to save space.

Actually, the occurrence graph tool has a data representation that is slightly more complex than the representation which we have described above. The marking of a page is not represented by a single page record. Instead it is represented by a set of page instance records – two for each page instance. The first record corresponds to those place instances that are independent of other place instances (i.e., belong to a place instance group with only one element). The second record corresponds to those place instances that depend on other place instances (via fusion sets and port assignments).

Practical experience shows that typical O-graphs use 1–2 KB of memory for each node. This could be significantly reduced by an increased coding of the markings (in particular of the multi-set representations). However, this would imply that the construction and analysis algorithms become slower, since they have to pack and unpack the multi-sets. Another possibility is to omit those parts of a marking which can be uniquely determined from others parts. For the data base system it would be sufficient to record Rec(M(Sent)), Rec(M(Received)), and Rec(M(Acknowledged)), where Rec is the function which maps each message (s,r) into the receiver r.

Key functions

From the above description of the search trees, it is clear that we want a set of key functions which map the different records into many different key values. However, there is a trade-off between this concern and the amount of time and space which we use to calculate and store key values. In practice, it turns out that it is often best to use a set of relatively simple key functions.

For multi-set records the occurrence graph tool offers three standard key functions. The first key function maps each multi-set into its size (e.g., K_1(2\`a+1\`c+1\`d+3\`g) = 7). The second key function maps each multi-set into its **coefficient multi-set** (e.g., K_2(2\`a+1\`c+1\`d+3\`g) = 2\`1+1\`2+1\`3). The third key function maps each multi-set into the **coefficient list** where the coefficients appear in the order determined by the order of their colours (e.g., $K_3$(2\`a+1\`c+1\`d+3\`g) = [2,1,1,3]). For K_2 and K_3 we only consider non-zero coefficients, and we encode the coefficient multi-set and the coefficient list into a single integer. For multi-sets with few coefficients the encoding is done in a bijective way. However, for large multi-sets it may be necessary to throw some of the coefficients away, because we operate with a fixed key length. Also for K_1 it is necessary to handle overflow problems – although they seldom appear in practice. The user can choose one of the three standard key functions, or he can construct his own.

The standard key function of a page instance record is calculated from the keys of the involved multi-set records. This is done by means of an appropriate encoding function. Analogously, the standard key function of a marking record is calculated from the keys of the involved page instance records. In both cases it is possible for the user to define his own key functions.

The best choice for a set of key functions depends upon the characteristics of the CP-net in question. For large O-graphs it is recommended to spend some time considering and evaluating the different possibilities. The choice of key functions may have quite a dramatic effect upon the time complexity of the O-graph algorithm.

Partial O-graphs

When the O-graph is infinite or too big to be constructed by the available computing power, it may still be of interest to construct a partial O-graph, i.e., a subgraph of the O-graph. The occurrence graph tool supports this by means of a set of stop criteria and a set of branching criteria.

The **stop criteria** allow the user to specify how large the constructed graph should be. As an example, it is possible to specify that the construction should finish when a certain amount of real time has been used or when a certain number of nodes have been constructed. It is also possible to specify a predicate function that checks each node as soon as it has been processed. In this way it is possible, e.g., to say that the O-graph constructor should stop when three dead markings have been found.

The **branching criteria** imply that we do not develop all the successor markings of a given node. As an example, the user may specify that he wants the construction to investigate at most three transition instances and at most two bindings (for each of the three transition instances). It is also possible to specify a predicate function which checks each node before any successors are calculated. In this way it is possible to say, e.g., that the O-graph constructor should ignore all nodes where the marking has a certain property.

When a partial graph has been constructed, by using a set of stop criteria or a set of branching criteria, it is later possible to continue the construction (with or without stop and branching criteria). The construction is always breadth first. This means the set *Waiting* in the algorithm of Prop. 1.4 is handled as a FIFO-queue. However, a depth first construction can be obtained, to a certain degree, by specifying a restrictive set of branching criteria.

The analysis of a partial O-graph can never give a total proof of the system correctness, since we can say nothing about those reachable markings and occurring steps that are not present in the O-graph. However, a partial O-graph often allows us to identify undesired properties of the modelled system. If a partial O-graph contains a processed node without successors we have found a dead marking. If a partial O-graph has an upper bound that is larger than expected, this will also be the case in the full O-graph. A full discussion of the proof rules for partial occurrence graphs is outside the scope of this book.

Graphical display of O-graphs

It is not possible to draw a large O-graph – nor would it be useful. However, even for a large O-graph, it often makes sense to display selected parts, e.g., the immediate surroundings of some interesting nodes/arcs – which may be found via a set of search functions to be introduced in the following subsections.

The occurrence graph tool supports the drawing of partial or full O-graphs, e.g., those shown in Figs. 1.2 and 1.3. The user specifies the nodes he wants the tool to display. This can be done by explicitly listing the desired nodes. It can also be done by asking for the successors or predecessors of some already displayed nodes. The new nodes are drawn with a default position and with default attributes. The objects of the O-graph are ordinary graphical objects, and this means the user can change the attributes. Usually, he will modify the positions as he adds new nodes to get a nice layout without too many crossing arcs.

The user determines the text strings to be displayed for the individual nodes and arcs. This is done by specifying two SML functions. One of these maps from nodes into text strings, while the other maps from arcs into text strings. Usually, the first function gives a condensed representation of the marking (of the node) and the second gives a condensed representation of the binding element (of the arc). Each text string is displayed in a region of the corresponding node/arc, and the region can be made visible/invisible by double clicking the node/arc. In Figs. 1.2 and 1.3 we have made all the regions visible. Each node region is positioned on top of the corresponding node, while each arc region is positioned next to the corresponding arc.

The above facilities make it fast to draw small O-graphs like Figs. 1.2 and 1.3 (or small parts of a large O-graph). First the user writes the two text functions (determining how he wants the markings and binding elements to be displayed). This is done by applying a set of predefined SML functions. Then the graph is drawn. The occurrence graph tool makes all the calculations and creates the necessary nodes, arcs, and regions. The user only adjusts the layout and determines the extent of the drawing.

The user can split the display of a large O-graph into a number of different pages, e.g., one for each strongly connected component. Then the system maintains a consistent representation of the connections between the individual pages.

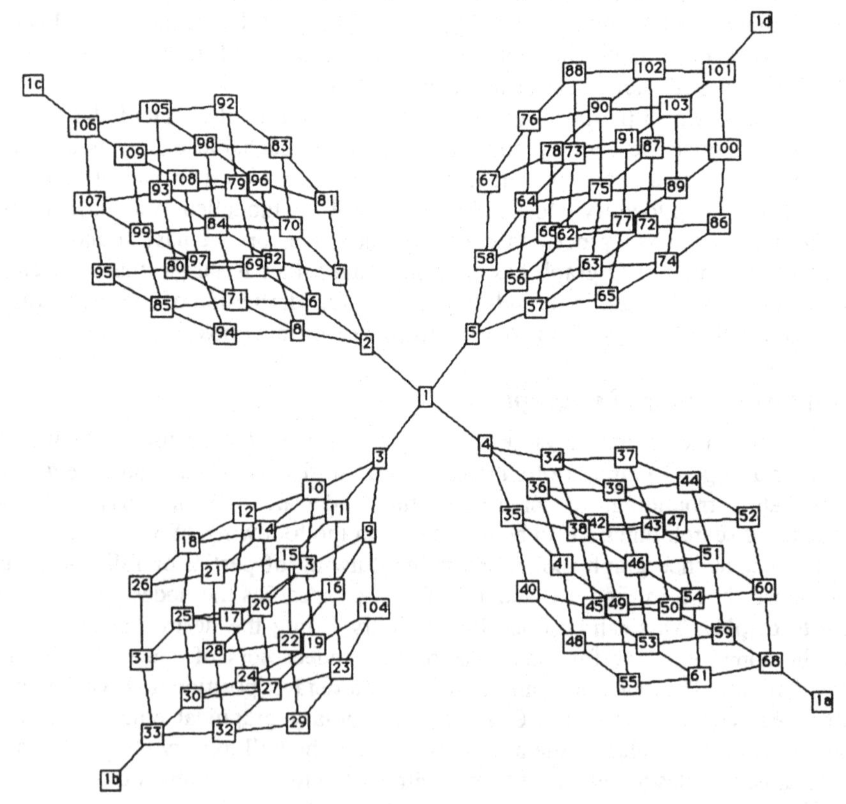

Fig. 1.10. O-graph for data base system with 4 managers (semi-automatic layout)

This works in a way that is similar to the port and socket nodes used in hierarchical nets.

We intend to add a semi-automatic layout routine to the occurrence graph tool. This means the system will be able to draw O-graphs like the one shown in Fig. 1.10. The layout routine is interactive. This means the user can move a node. Then the system will readjust the remaining nodes. For more details, see the bibliographical remarks.

Analysis of O-graphs

Most of the proof rules presented in Sect. 1.4 are based on the possibility of making a brute-force search in which every node or every arc of the O-graph (or one of its strongly connected components) is visited to gather some information. To perform such searches we use two predefined functions, **SearchNodes** and **SearchArcs**. Each of the two functions takes six different arguments, which are supplied to specify the details of the search and the details of the involved computations. For SearchNodes these arguments are as follows:

Search area	This argument specifies the part of the graph which should be searched. It is often all the nodes of the O-graph, but it may also be any other subset of nodes, e.g., those belonging to a strongly connected component.
Predicate function	This argument specifies a function. It maps each node into a boolean value. Those nodes which evaluate to false are ignored; the others take part in the further analysis – as described below.
Search limit	This argument specifies an integer. It tells us how many times the predicate function may evaluate to true before we terminate the search. The search limit may be ∞ (which denotes infinity). This means we always search through the entire search area.
Evaluation function	This argument specifies a function. It maps each node into a value, of some type A. It is important to notice that the evaluation function is only used at those nodes (of the search area) for which the predicate function evaluates to true.
Start value	This argument specifies a constant, of some type B.
Combination function	This argument specifies a function. It maps from A×B into B, and it describes how each individual result (obtained by the evaluation function) is combined with the prior results.

It should be noted that the predicate function, the evaluation function, and the combination function are all written by the user. This means they can be arbitrarily complex.

The SearchNodes function works as described by the following Pascal-style pseudo-code. When the function terminates it returns the value of *Result*:

```
SearchNodes (Area, Pred, Limit, Eval, Start, Comb)
begin
    Result := Start; Found := 0
    for all n∈Area do
        if Pred(n) then
        begin
            Result := Comb(Eval(n), Result)
            Found := Found + 1
            if Found = Limit then stop for-loop
        end
    end
end.
```

The SearchNodes function may seem a bit complicated. But it is also extremely general and powerful. As an example, we can use SearchNodes to implement the proof rule in Prop. 1.15 (i), i.e., to verify whether there are any dead markings. Then we use the following arguments:

Search area	V
Predicate function	fun Pred(n) = (length(OutArcs(n)) = 0)
Search limit	10
Evaluation function	fun Eval(n) = n
Start value	[ ]
Combination function	fun Comb(new,old) = new::old

The predicate function uses the predefined function *OutArcs* to get a list of all the output arcs. If the length of this list is zero there are no successors, and thus we have a dead marking. The evaluation function maps a node into itself, i.e., into the unique node number. The combination function adds each new dead marking to the list of those which we have previously found. With these arguments SearchNodes returns a list with at most 10 dead markings. If the list is empty there is no dead marking. If the length is less than 10, the list contains all the dead markings.

When we deal with a partial O-graph in which some of the nodes have not yet been processed, it is better to have a slightly more complex predicate function. We then avoid those nodes we have not yet processed. This is done by using the predefined function *NodeProcessed* that indicates whether the node has been processed or not:

Pred. func.	fun Pred(n) = (length(OutArcs(n)) = 0 $\wedge$ NodeProcessed(n))

When SearchNodes (with the modified predicate function) returns a non-empty list, we know that all the nodes in the list correspond to dead markings. When SearchNodes returns an empty list, we only know that all the nodes which the occurrence constructor has processed are non-dead. However, since the O-graph is only partial, there are reachable markings that we have not yet found or not yet processed. For such nodes we have no information whether they are dead or not.

As a second example, we may use SearchNodes to implement the proof rule in Prop. 1.13 (iii), i.e., to find the best upper multi-set bound for a given place instance $p \in PI$. This is done by using the following arguments:

Search area	V
Predicate function	fun Pred(n) = true
Search limit	∞
Evaluation function	fun Eval(n) = Mark(p)(n)
Start value	$\emptyset$
Combination function	The max function for multi-sets (defined in Sect. 1.4)

The evaluation function uses a predefined function *Mark* to get the marking of the place instance p at the node n. It should be obvious that we can find all the other kinds of bounds in a similar way (i.e., lower bounds, integer bounds, and bounds determined by a set of binding elements or a function).

SearchArcs is very similar to SearchNodes. It takes the same six arguments. However, it searches through the arcs instead of the nodes, and this means the search area is a set of arcs, while the predicate function and the evaluation function take an arc as argument.

Strongly connected components

The occurrence graph tool can construct SCC-graphs. This is done by means of a standard algorithm from graph theory. The algorithm has a linear time complexity, $O(\max(|V|,|A|))$. For more details, see the bibliographical remarks. The constructed SCC-graph can be displayed graphically – like the O-graph.

It is possible to make a brute-force search of all nodes in an SCC-graph, i.e., all the strongly connected components. This is done by means of the predefined function **SearchComponents**. It works in a similar way to SearchNodes and SearchArcs and it takes the same six arguments. The search area is a subset of SCC, while the predicate function and the evaluation function take a member of SCC as argument.

As an example, we use SearchComponents to implement the proof rule in Prop. 1.15 (ii). This means we use the SCC-graph to verify whether there are any dead markings. To do this we use the following arguments:

Search area	SCC
Predicate function	fun Pred(c) = Terminal(c) $\wedge$ Trivial(c)
Search limit	∞
Evaluation function	fun Eval(c) = c
Start value	[]
Combination function	fun Comb(new,old) = new::old

The predicate function uses two predefined functions to test whether c is terminal and trivial. The evaluation function maps a component into itself, i.e., into the unique component number. The combination function adds each new component to the list of those we have previously found. This means we end up with a list that contains all the dead markings (or more precisely, a list of the corresponding components).

As a second and more complicated example, we use SearchComponents to implement the proof rule in Prop. 1.15 (iv), i.e., to verify whether a given set of binding elements is live or not. To do this we simply have to check whether each terminal strongly connected component contains at least one element from X. This is done by using SearchComponents with the following arguments:

<table>
<tr><td>Search area</td><td colspan="2">SCC</td></tr>
<tr><td>Predicate function</td><td colspan="2">fun Pred(c) = Terminal(c)</td></tr>
<tr><td>Search limit</td><td colspan="2">∞</td></tr>
<tr><td rowspan="7">Evaluation function</td><td colspan="2">fun Eval(c) = SearchArcs(...)
with the following arguments:</td></tr>
<tr><td>Search area</td><td>c</td></tr>
<tr><td>Pred. func.</td><td>fun Pred(a) = (Bind(a) $\in$ X)</td></tr>
<tr><td>Search limit</td><td>1</td></tr>
<tr><td>Eval. func.</td><td>fun Eval(a) = true</td></tr>
<tr><td>Start value</td><td>false</td></tr>
<tr><td>Comb. func.</td><td>fun Comb(new,old) = new</td></tr>
<tr><td>Start value</td><td colspan="2">true</td></tr>
<tr><td>Combination function</td><td colspan="2">fun Comb(new,old) = new $\wedge$ old</td></tr>
</table>

The predicate function for SearchArcs uses a predefined function *Bind* to get the binding element of the arc a. To understand the combination function for

SearchArcs, it should be remembered that the evaluation function is only used for those arcs that satisfy the predicate function.

Standard queries

Above we have illustrated how SearchNodes, SearchArcs, and SearchComponents can be used to make an automatic verification of the dynamic properties of a CP-net.

For each of the examples above, we have shown the six arguments of the search functions, using a notation which is similar to the actual SML code used by the occurrence graph tool. Each call of a search function typically uses 2–5 lines of SML code, and can be written without a deep knowledge of SML. However, most users do not need to write arguments for the search functions. Instead they use a number of predefined functions that implement standard queries by means of the search functions and based upon the proof rules in Sect. 1.4. Some of these standard queries can be used directly, while others must be slightly modified. As an example, there is a standard query which verifies whether a transition is live. This query can be used directly. The only thing the user has to do is to specify which transition he wants to consider. There are also standard queries to find bounds for an arbitrary set of token elements $X \subseteq TE$. To use these queries the user has to write a function mapping each marking M into $(M | X)$. This is a rather straightforward task and does not require any deep knowledge of SML.

In addition to the standard queries, the occurrence graph tool allows the more experienced user to program his own queries by specifying non-standard arguments to one of the three search functions. This means the tool can be used to answer a very large set of widely different queries.

Bibliographical Remarks

An O-graph contains a node for each reachable marking and an arc for each occurring binding element. Unfortunately, the literature has many different names for such graphs. Some authors talk about state spaces, others talk about reachability graphs or reachability trees. There is no substantial difference between these concepts. I have chosen to talk about occurrence graphs, because this name conveys the fact that the O-graph represents all the possible occurrence sequences, i.e., both the reachable markings and the occurring steps. The other names are slightly more biased towards the markings. Occurrence graphs should not be confused with occurrence nets – the two concepts have nothing to do with each other.

O-graphs are widely used and have been known for many years, both for Petri nets and for arbitrary transition systems. Directed paths, strongly connected components, and SCC-graphs are standard concepts from graph theory and can be found in most introductory text books, e.g., [2] and [19]. Most of the proof rules presented in Sect. 1.4 are generalisations of proof rules that for many years have been widely known for PT-nets (and other kinds of transition system). However, as far as we know, the proof rules for the fairness properties are novel.

The occurrence graph tool is described in [25]. It computes the strongly connected components by means of Tarjan's algorithm, which can be found in [2] and [19]. The layout algorithm used to produce Fig. 1.10 has been taken from [3]. It is based upon an algorithm presented in [27].

Exercises

Exercise 1.1.
Consider the simplified resource allocation system from Fig. 1.1 and the O-graph in Fig. 1.2.

(a) Check that the O-graph is correct. This can be done by means of an occurrence graph tool (or a CPN simulator).

(b) Use the O-graph and the proof rules in Prop. 1.13 to investigate the boundedness properties of the simplified resource allocation system.

(c) Use the O-graph and the proof rules in Prop. 1.14 to investigate the home properties of the simplified resource allocation system.

(d) Use the O-graph and the proof rules in Prop. 1.15 to investigate the liveness properties of the simplified resource allocation system.

(e) Use the O-graph and the proof rules in Prop. 1.16 to investigate the fairness properties of the simplified resource allocation system.

(f) Compare the results in (b)–(e) with the dynamic properties postulated for the non-simplified resource allocation system at the end of Sects. 4.2–4.5 of Vol. 1.

Exercise 1.2.
Consider the data base system from Sect. 1.3 of Vol. 1 and the O-graph in Fig. 1.4.

(a) Check that the O-graph is correct. This can be done by means of an occurrence graph tool (or a CPN simulator).

(b) Use the O-graph and the proof rules in Prop. 1.13 to verify the upper and lower bounds postulated at the end of Sect. 4.2 of Vol. 1.

(c) Use the O-graph and the proof rules in Prop. 1.14 to verify the home properties postulated at the end of Sect. 4.3 of Vol. 1.

(d) Use the O-graph and the proof rules in Prop. 1.15 to verify that SM and RA are strictly live, while RM and SA are live.

(e) Use the O-graph and the proof rules in Prop. 1.16 to verify the fairness properties postulated at the end of Sect. 4.5 of Vol. 1.

Exercise 1.3:

Consider the philosopher system from Sect. 1.6 and the O-graph in Fig. 1.7.

(a) Check that the O-graph is correct. This can be done by means of an occurrence graph tool (or a CPN simulator).

(b) Use the O-graph and the proof rules in Prop. 1.13 to investigate the boundedness properties of the philosopher system.

(c) Use the O-graph and the proof rules in Prop. 1.14 to investigate the home properties of the philosopher system.

(d) Use the O-graph and the proof rules in Prop. 1.15 to investigate the liveness properties of the philosopher system.

(e) Use the O-graph and the proof rules in Prop. 1.16 to investigate the fairness properties of the philosopher system.

(f) Repeat (b)–(e) with $|PH| = |CS| = 7$ and with $|PH| = |CS| = 9$ (if possible). Now you have to construct the O-graph yourself by means of an occurrence graph tool.

Exercise 1.4.

Consider the telephone system from Sect. 3.2 of Vol. 1. This exercise is only worth attempting if you have access to an occurrence graph tool. Even though the net is rather small, it will take too long to produce the O-graphs if this has to be done manually.

(a) Construct an O-graph for the telephone system with $|U| = 3$. What is the size of the O-graph?

(b) Use the O-graph and the proof rules in Prop. 1.13 to verify the upper and lower bounds postulated at the end of Sect. 4.2 of Vol. 1.

(c) Use the O-graph and the proof rules in Prop. 1.14 to verify the home properties postulated at the end of Sect. 4.3 of Vol. 1.

(d) Use the O-graph and the proof rules in Prop. 1.15 to verify the liveness properties postulated at the end of Sect. 4.4 of Vol. 1. Investigate whether any of the transitions are strictly live.

(e) Use the O-graph and the proof rules in Prop. 1.16 to verify the fairness properties postulated at the end of Sect. 4.5 of Vol. 1.

(f) Repeat (a)–(e) with $|U| = 4$ and with $|U| = 5$ (if possible).

Exercise 1.5.

Consider the process control system from Exercise 4.6 of Vol. 1. This exercise is only worth attempting if you have access to an occurrence graph tool. Even though the net is rather small, it will take too long to produce the O-graphs if this has to be done manually.

(a) Construct an O-graph for the process control system with |PROC| = |RES| = 2. What is the size of the O-graph?

(b) Use the O-graph and the proof rules in Prop. 1.13 to investigate the boundedness properties of the process control system.

(c) Use the O-graph and the proof rules in Prop. 1.14 to investigate the home properties of the process control system.

(d) Use the O-graph and the proof rules in Prop. 1.15 to investigate the liveness properties of the process control system.

(e) Use the O-graph and the proof rules in Prop. 1.16 to investigate the fairness properties of the process control system.

(f) Repeat (a)–(e) with |PROC| = |RES| = 3 and with |PROC| = |RES| = 4 (if possible).

Exercise 1.6.
Consider the ring network from Sect. 3.1 of Vol. 1. This exercise is only worth attempting if you have access to an occurrence graph tool. Even though the net is rather small, it will take too long to produce the O-graphs if this has to be done manually.

(a) Modify the ring network in the following way:
   - Remove *PackNo* and the three reporting places *SentExt*, *RecInt* and *RecExt*.
   - Modify *NewPack* as follows. Each of the four sites has a fixed set of message buffers: 1..NoOfBuffers. A site can only send a message when one of its message buffers is free. All messages {se=s, re=r, no=i} satisfy s ≠ r.
   - Modify *Receive* as follows. When a message {se=s, re=r, no=i} reaches its final destination r, an acknowledgement message {se=r, re=s, no=~i} is generated and put on *Package*. When the acknowledgement message reaches its final destination s, the message buffer i becomes free.

(b) Make a simulation of the modified CP-net.

(c) Construct an O-graph for the ring network with NoOfBuffers = 2. What is the size of the O-graph?

(d) Use the O-graph and the proof rules in Prop. 1.13 to investigate the boundedness properties of the modified ring network.

(e) Use the O-graph and the proof rules in Prop. 1.14 to investigate the home properties of the modified ring network.

(f) Use the O-graph and the proof rules in Prop. 1.15 to investigate the liveness properties of the modified ring network.

(g) Use the O-graph and the proof rules in Prop. 1.16 to investigate the fairness properties of the modified ring network.

(h) Repeat (c)–(g) with NoOfBuffers = 3 and with NoOfBuffers = 4 (if possible).

Exercise 1.7.
Consider Prop. 1.16, which deals with proof rules for the fairness properties.

(a) Fill in the details of the proofs for properties (ii)–(iii).

(b) Fill in the details of the proofs for properties (iv)–(vi).

Exercise 1.8.
Consider the O-graphs for the philosopher system in Sect. 1.6.

(a) Prove that the number of nodes grows as a Fibonacci sequence, i.e., that $N_O(n) = N_O(n-1) + N_O(n-2)$ where $N_O(2) = 3$ while $N_O(3) = 4$.

(b) Try to prove that the number of arcs is $A_O(n) = 2 * n * F(n)$ where $F(n)$ is the Fibonacci sequence determined by $F(2) = F(3) = 1$.

Chapter 2

Occurrence Graphs with Equivalence Classes

This chapter deals with a more complex kind of occurrence graphs, called occurrence graphs with equivalence classes (OE-graphs). To obtain an OE-graph the user has to specify an equivalence relation for the set of markings and an equivalence relation for the set of binding elements. Together these two equivalence relations are called an equivalence specification. The OE-graph has a node for each equivalence class of markings that contains a reachable marking. Analogously, the OE-graph has an arc, between two nodes, for each equivalence class of binding elements containing a binding element that is enabled in a marking of the source node and leads to a marking of the destination node.

OE-graphs are often much smaller than the corresponding O-graphs. Nevertheless, OE-graphs can still be used to verify many of the dynamic properties defined in Vol. 1, i.e., the reachability, boundedness, home, liveness, and fairness properties. To make this possible the equivalence specification must be chosen in such a way that it is consistent/compatible with the enabling and occurrence rule of the CP-net (for more details, see Sect. 2.2).

The construction of OE-graphs and the associated verification of dynamic properties can be fully automated. This means OE-graphs provide a very powerful and relatively easy-to-use method for analysing the properties of a given CP-net.

Section 2.1 contains an informal introduction to OE-graphs. This is done by means of the resource allocation system, i.e., the same system as we used to introduce O-graphs. Section 2.2 contains the formal definition of OE-graphs. Section 2.3 contains a number of detailed proof rules, i.e., propositions allowing us to prove (or disprove) different CP-net properties by inspection of different OE-graph properties. The proof rules are modifications of the proof rules for O-graphs. Section 2.4 illustrates that the sizes of OE-graphs may grow much more slowly than the sizes of O-graphs – when the sizes of the involved colour sets increase. This is shown by considering the data base system. Section 2.5 gives an additional example of OE-graphs. Section 2.6 extends OE-graphs by adding a set of node and arc labels giving us information about the behaviour of individual markings and binding elements. Finally, Sect. 2.7 discusses how the construction and analysis of OE-graphs are supported by computer tools.

2.1 Introduction to OE-graphs

It is often the case that a CP-net has some markings which, for certain purposes, are so alike that we want to ignore the differences between them. As an example, consider Chap. 4 of Vol. 1 in which we ignored the cycle counters of the resource allocation system when analysing the dynamic properties. This was done by defining an equivalence relation $\approx_A$ on the set of all binding elements BE (see Sect. 4.4 of Vol. 1) and by defining a projection function that maps each multi-set in P_{MS} into a multi-set in U_{MS} – by throwing away the second component of each token colour (see Sect. 4.2 of Vol. 1). From the projection function it is straightforward to define an equivalence relation $\approx_M$ on the set of all markings $\mathbb{M}$. We simply say that two markings M_1 and M_2 are equivalent iff the following two properties are satisfied:

$$\forall p \in \{R,S,T\}: \ M_1(p) = M_2(p)$$
$$\forall p \in \{A,B,C,D,E\}: \ PR_1(M_1(p)) = PR_1(M_2(p)).$$

This equivalence relation was already used implicitly in Sect. 4.3 of Vol. 1, where we showed that the equivalence class $[M_0]$ is a home space for the resource allocation system.

When the markings in an equivalence class are alike, it may be possible to represent all of them by a single node. Analogously, when the binding elements in an equivalence class are alike, it may be possible to represent all of them by a single arc. This means each node of the occurrence graph represents an equivalence class of markings [M] while each arc represents a triple $([M_1],[b],[M_2])$ where [b] is an equivalence class of binding elements "leading" from markings of $[M_1]$ to markings of $[M_2]$.

It is indeed possible to construct an occurrence graph in the way described above. Such a graph is called an **occurrence graph with equivalence classes** or an **OE-graph**. Now the next question is whether OE-graphs are useful – in the sense that we can use them to investigate the dynamic properties of CP-nets. Again the answer is positive – but only under certain circumstances. We need to choose the two equivalence relations $\approx_M$ (on the set of markings) and $\approx_{BE}$ (on the set of binding elements) in such a way that they match the behaviour of the CP-net. When two markings M_1 and M_1^* are equivalent we require that, for each step $M_1 [b\rangle M_2$, there exists a step $M_1^* [b^*\rangle M_2^*$ where b^* is equivalent to b and M_2^* is equivalent to M_2. Without this property, we would not know whether the OE-graph should have an arc $([M_1],[b],[M_2])$. By focusing on M_1 we ought to have the arc (because $M_1 [b\rangle M_2$ is a step). However, by focusing on M_1^* we should not have the arc (because M_1^* has no step of the form $M_1^* [b^*\rangle M_2^*$).

When we have defined the equivalence relations $\approx_M$ and $\approx_{BE}$ so that they satisfy the property described above, we say that we have an **equivalence specification** that is **consistent** with the behaviour of the CP-net. It can then be shown that the corresponding OE-graph can be used to investigate the dynamic properties of the CP-net. This is done by means of a set of proof rules, which are modifications of the proof rules defined for O-graphs in Sect. 1.4. In Sect. 2.2 we give a formal definition of OE-graphs for consistent equivalence specifications. The proof rules for OE-graphs will be defined in Sect. 2.3. In Sects. 2.2 and 2.3 we also deal with a weaker concept called a **compatible** equivalence specification. We show that, for compatible equivalence specifications, it also makes sense to generate and analyse OE-graphs.

Let us now consider again the resource allocation system from Fig. 1.7 of Vol. 1, i.e., the version with cycle counters. First we must verify that the two equivalence relations $\approx_M$ and $\approx_{BE}$ constitute a consistent equivalence specification. This is easy. Let M_1 be a marking with a step $M_1 [b\rangle M_2$ and let $M_1{}^*$ be a marking which is equivalent to M_1. We then know that $M_1{}^*$ and M_1 are identical except for the cycle counters in the second component of tokens on places A–E. We also know that the binding element b is in the form $b = (Tn, \langle x=v, i=w \rangle)$ where $n \in 1..5$, $v \in \{p,q\}$ and $w \in \mathbb{Z}$. Let $p \in \{A,B,C,D,E\}$ be input place of transition Tn. Since b is enabled in M_1 we conclude that $M_1(p)$ must contain a token of the form (v,w). From $M_1{}^* \approx_M M_1$ we then know that $M_1{}^*(p)$ has a token of the form (v,w*), and from this we conclude that the binding element $b^* = (Tn, \langle x=v, i=w^* \rangle)$ is enabled in $M_1{}^*$. It is easy to see that b* is equivalent to b and that the occurrence of b* in $M_1{}^*$ gives us a marking $M_2{}^*$ which is equivalent to M_2.

Next we construct the OE-graph. When we draw it, each node is inscribed by a text string representing an equivalence class of markings. Analogously, each arc is inscribed by a text string representing an equivalence class of binding elements. Two markings of the resource allocation system are equivalent iff they are identical when we ignore the cycle counters. Hence it is obvious to represent the equivalence classes by showing the markings which we get when we remove the cycle counters. Something similar is done for the equivalence classes of binding elements. With this representation it can be seen that Fig. 1.2 in Sect. 1.1 is an OE-graph for the resource allocation system.

From the OE-graph in Fig. 1.2 it is possible to verify most of the dynamic properties postulated for the resource allocation system in Chap. 4 of Vol. 1 (see Exercise 2.1). The existence of a node [M] tells us that some of the markings in [M] are reachable – but not necessarily all of them. Those with negative cycle counters are not reachable (but this cannot be deduced from the OE-graph). The existence of an arc [b], from a node $[M_1]$ to a node $[M_2]$, tells us that for each marking $M_1{}^* \in [M_1]$ there exist a binding element $b^* \in [b]$ and a marking $M_2{}^* \in [M_2]$ such that $M_1{}^* [b^*\rangle M_2{}^*$.

From the discussion above, it may seem that we have not achieved very much by using equivalence classes. However, this is not at all correct. It should be remembered that Fig. 1.2 is an O-graph for the simplified resource allocation system in Fig. 1.1, while it is an OE-graph for the non-simplified resource allocation system in Fig. 1.7 of Vol. 1. Thus, we have been able to construct a finite and relatively small OE-graph for a CP-net where the O-graph is infinite. It can be argued that, for the resource allocation system, we can achieve a similar effect by removing the cycle counters before we start the analysis of the system. This is correct. However, OE-graphs can be used in a lot of situations where it is not obvious how to simplify the original CP-net. By means of OE-graphs we can perform occurrence graph analysis of CP-nets that we would be unable to handle with O-graphs, because the O-graphs would be infinite or too large to construct in practice. In Sect. 2.4 we construct OE-graphs for the data base system, and we show that the number of nodes and the number of arcs are both of the order $O(n^2)$, where n is the number of data base managers. This means the size of the OE-graph grows much more slowly than the size of the O-graph, for which the number of nodes is of order $O(n*3^n)$ while the number of arcs is of order $O(n^2*3^n)$. The OE-graph of the data base system is of polynomial space complexity, while the O-graph is of exponential space complexity.

2.2 Formal Definition of OE-graphs

In this section we define OE-graphs of CP-nets, and we also give an algorithm to construct OE-graphs. First we define equivalence specifications.

Definition 2.1: An **equivalence specification** for a CP-net is a pair $(\approx_M, \approx_{BE})$ where $\approx_M$ is an equivalence relation on $\mathbb{M}$ while $\approx_{BE}$ is an equivalence relation on BE.

We use $M_\approx$ to denote the set of all equivalence classes for $\approx_M$ and we use $BE_\approx$ to denote the set of all equivalence classes for $\approx_{BE}$. Two pairs $(b,M), (b^*,M^*) \in BE \times \mathbb{M}$ are said to be equivalent to each other, $(b,M) \approx (b^*,M^*)$, iff $b \approx_{BE} b^*$ and $M \approx_M M^*$.

For a set of markings $X \subseteq \mathbb{M}$, we use [X] to denote all the markings that are equivalent to a marking from X. Analogously, for a set of equivalence classes $Y \subseteq M_\approx$, we use [Y] to denote all the markings that belong to one of the equivalence classes in Y:

$$[X] = \{M \in \mathbb{M} \mid \exists x \in X: M \approx_M x\}$$
$$[Y] = \{M \in \mathbb{M} \mid \exists y \in Y: M \in y\}.$$

An analogous notation is used for a set $X \subseteq BE$, a set $Y \subseteq BE_\approx$ and a set $X \subseteq BE \times \mathbb{M}$.

It is possible to define many different equivalence specifications. As one extreme, we may choose $\approx_M$ and $\approx_{BE}$ so restrictively that a marking (or a binding element) is only equivalent to itself. As another extreme, we may choose $\approx_M$ so that everything becomes equivalent to each other. The choice of the equivalence specification will often have a significant impact upon the set of properties we can investigate (see the proof rules in Sect. 2.3). Hence it is important to choose the equivalence specification with some care.

To be useful for the construction of OE-graphs we require the equivalence specification to be chosen in such a way that equivalent markings/binding elements are known to have similar behavioural properties. To achieve this we define consistency and compatibility of equivalence specifications. The definition of consistency uses the notation $Next(M_1)$, which we introduced immediately above Prop. 1.4. The definition of compatibility uses the notation $Next_\tau(M_1)$, which we will introduce and explain after Def. 2.2. In Sect. 2.7 we discuss how the check of consistency/compatibility can be supported by computer tools.

Definition 2.2: An equivalence specification is **consistent** iff the following property is satisfied, for all $M_1, M_2 \in [[M_0\rangle]$:

(i) $M_1 \approx_M M_2 \Rightarrow [Next(M_1)] = [Next(M_2)]$.

An equivalence specification is **compatible** iff the following property is satisfied, for all $M_1, M_2 \in [[M_0\rangle]$:

(ii) $M_1 \approx_M M_2 \Rightarrow [Next_\tau(M_1)] = [Next_\tau(M_2)]$.

Property (i) guarantees that two equivalent markings M_1 and M_2 have a similar behaviour. For each step $M_1[b\rangle M$ we require that there exists a step $M_2[b^*\rangle M^*$ such that $(b,M) \approx (b^*, M^*)$. This is a rather strong demand, which requires a very close relationship between the direct successors of M_1 and the direct successors of M_2.

Property (ii) is a weaker property than (i). Now we ignore those steps where the end marking belongs to the same equivalence class as the start marking. We shall use τ to denote a sequence of such steps, and we shall consider the steps to be **non-observable** – because the marking remains inside the same equivalence class. This idea is formalised as follows.

Definition 2.2 requires that (i) and (ii) be satisfied for all markings in $[[M_0\rangle]$. In practice, we will often show that the two properties are satisfied for all markings in $\mathbb{M}$. Then we can verify the consistency properties without constructing $[[M_0\rangle]$.

For two markings $M_1, M \in \mathbb{M}$ and a binding element $b \in BE$, we use $M_1[\tau\ b\rangle M$ to denote the fact that there exists a finite occurrence sequence:

$$M_1[b_1\rangle M_2[b_2\rangle M_3 \ldots M_n[b_n\rangle M_{n+1}[b\rangle M$$

such that $n \in \mathbb{N}$, $b_i \in BE$ for all $i \in 1..n$ and $M_i \approx_M M_1$ for all $i \in 1..n+1$, while $M \not\approx_M M_1$. Moreover, we define $Next_\tau(M_1)$ as follows:

$$Next_\tau(M_1) = \{(b,M) \in BE \times \mathbb{M} \mid M_1[\tau\ b\rangle M\}.$$

It is straightforward to prove that each consistent equivalence specification is compatible. This is done by induction over the number of τ steps in the occurrence sequences considered in the definition of $Next_\tau$.

Definition 2.3: Let a CP-net and a consistent equivalence specification $(\approx_M, \approx_{BE})$ be given. The **occurrence graph with equivalence classes**, also called the **OE-graph**, is the directed graph $OEG = (V, A, N)$ where:

(i) $V = \{C \in \mathbb{M}_\approx \mid C \cap [M_0\rangle \neq \emptyset\}$.
(ii) $A = \{(C_1,B,C_2) \in V \times BE_\approx \times V \mid \exists (M_1,b,M_2) \in C_1 \times B \times C_2: M_1[b\rangle M_2\}$.
(iii) $\forall a=(C_1,B,C_2) \in A$: $N(a) = (C_1,C_2)$.

For a compatible equivalence specification we replace $M_1[b\rangle M_2$ by $M_1[\tau\ b\rangle M_2$.

The OE-graph has a node for each equivalence class [M] containing a marking $M^* \in [M]$ such that $M^* \in [M_0\rangle$. Analogously, the OE-graph has an arc for each triple of equivalence classes $([M_1],[b],[M_2])$ containing elements $M_1^* \in [M_1]$, $b^* \in [b]$ and $M_2^* \in [M_2]$ such that $M_1^* [b^*\rangle M_2^*$. The source node of the arc is the equivalence class of the start marking of the step, while the destination node is the equivalence class of the end marking. Some examples of OE-graphs are shown in Sect. 2.4.

From Def. 2.3 it is easy to verify that the graph obtained by using $M_1 [\tau b\rangle M_2$ in (ii) is identical to the graph obtained by using $M_1 [b\rangle M_2$ – except that the former contains no arcs for which the source is identical to the destination. In particular, this means the OE-graph of a compatible equivalence specification never has arcs for which the source is identical to the destination.

The consistency and compatibility properties in Def. 2.2 guarantee that there is a close relationship between the directed paths of the OE-graph and the occurrence sequences of the CP-net:

Proposition 2.4: For a consistent equivalence specification, the OE-graph satisfies the following properties:

(i) Each finite occurrence sequence:

$$M_1 [b_1\rangle M_2 [b_2\rangle M_3 \ldots M_n [b_n\rangle M_{n+1}$$

where $M_1 \in [M_0\rangle$ and $b_i \in BE$ for all $i \in 1..n$, has a **matching directed path**:

$$[M_1]\ ([M_1],[b_1],[M_2])\ [M_2]\ ([M_2],[b_2],[M_3])\ [M_3] \ldots$$
$$\ldots [M_n]\ ([M_n],[b_n],[M_{n+1}])\ [M_{n+1}].$$

(ii) Each finite directed path:

$$C_1\ (C_1,B_1,C_2)\ C_2\ (C_2,B_2,C_3)\ C_3 \ldots C_n\ (C_n,B_n,C_{n+1})\ C_{n+1}$$

has, for *each* marking $M_1 \in C_1$, a **matching occurrence sequence**:

$$M_1 [b_1\rangle M_2 [b_2\rangle M_3 \ldots M_n [b_n\rangle M_{n+1}$$

where $M_i \in C_i$ for all $i \in 2..n+1$ and $b_i \in B_i$ for all $i \in 1..n$.

Similar properties are satisfied for infinite occurrence sequences and infinite directed paths.

For a compatibility equivalence specification, properties (i) and (ii) are satisfied when we replace all appearances of $[b_i\rangle$ by $[\tau\, b_i\rangle$.

Proof: The two properties are proved by means of induction over the length of the occurrence sequence/directed path. The proof is similar to the proof of Prop. 1.6.

Property (i): When the length of the occurrence sequence is zero, property (i) follows directly from Def. 2.3 (i). Next, assume that property (i) is satisfied when the length of the occurrence sequence is n–1, and assume that we have a finite occurrence sequence:

$$M_1\,[b_1\rangle\, M_2\,[b_2\rangle\, M_3\ \ldots\ M_n\,[b_n\rangle\, M_{n+1}$$

where $M_1 \in [M_0\rangle$ and $b_i \in BE$ for all $i \in 1..n$. From the inductive hypothesis, it follows that there exists a finite directed path:

$$[M_1]\ ([M_1],[b_1],[M_2])\ [M_2]\ ([M_2],[b_2],[M_3])\ [M_3]\ \ldots$$
$$\ldots\ [M_{n-1}]\ ([M_{n-1}],[b_{n-1}],[M_n])\ [M_n].$$

From the above occurrence sequence we see that M_{n+1} is reachable from $M_1 \in [M_0\rangle$. Hence M_{n+1} is reachable from M_0, and from Def. 2.3 (i) we know that $[M_{n+1}]$ belongs to V. From Def. 2.3 (ii) and the step $M_n\,[b_n\rangle\, M_{n+1}$ we conclude that $([M_n],[b_n],[M_{n+1}])$ is an arc of the OE-graph. This means that:

$$[M_1]\ ([M_1],[b_1],[M_2])\ [M_2]\ ([M_2],[b_2],[M_3])\ [M_3]\ \ldots$$
$$\ldots\ [M_n]\ ([M_n],[b_n],[M_{n+1}])\ [M_{n+1}]$$

is a finite directed path, as required.

Property (ii): When the length of the directed path is zero, property (ii) is trivially satisfied. Next, assume that property (ii) is satisfied when the length of the directed path is n–1, and assume that we have a finite directed path:

$$C_1\ (C_1,B_1,C_2)\ C_2\ (C_2,B_2,C_3)\ C_3\ \ldots\ C_n\ (C_n,B_n,C_{n+1})\ C_{n+1}.$$

From the inductive hypothesis, it follows that, for *each* marking $M_1 \in C_1$, there exists a finite occurrence sequence:

$$M_1\,[b_1\rangle\, M_2\,[b_2\rangle\, M_3\ \ldots\ M_{n-1}\,[b_{n-1}\rangle\, M_n$$

where $M_i \in C_i$ for all $i \in 2..n$ and $b_i \in B_i$ for all $i \in 1..n{-}1$. From Def. 2.3 (ii) and the arc (C_n,B_n,C_{n+1}) we conclude that there exists a step $M_n{}^*\,[b_n{}^*\rangle\, M_{n+1}{}^*$ of the CP-net such that $(M_n{}^*,b_n{}^*,M_{n+1}{}^*) \in C_n \times B_n \times C_{n+1}$. From Def. 2.2 (i) we then know that there exists a step $M_n\,[b_n\rangle\, M_{n+1}$ where $(b_n,M_{n+1}) \in B_n \times C_{n+1}$. This means that:

$$M_1\,[b_1\rangle\, M_2\,[b_2\rangle\, M_3\ \ldots\ M_n\,[b_n\rangle\, M_{n+1}$$

is a finite occurrence sequence with the requested properties. □

Below we give an abstract algorithm to construct the OE-graph for a consistent or compatible equivalence specification. The algorithm uses the same notation as the O-graph algorithm in Prop. 1.4. *Waiting* is a set of nodes, i.e., a set of equivalence classes of markings.

Proposition 2.5: The following algorithm constructs the OE-graph for a consistent equivalence specification. The algorithm halts iff V is finite and $Next(M_1)$ is finite for all selected markings M_1. When only the second of these properties is satisfied, the algorithm continues forever, producing a larger and larger subgraph of the labelled OE-graph.

```
Waiting := Ø
Node([M0])
repeat
    select a marking M1 ∈ [Waiting]
    for all (b,M2) ∈ Next(M1) do
    begin
        Node([M2])
        Arc([M1],[b],[M2])
    end
    Waiting := Waiting - {[M1]}
until Waiting = Ø.
```

For a compatible equivalence specification we replace Next by $Next_\tau$.

Proof: The proof is similar to the proof of Prop. 1.4. The main difference is that we now have to convince ourselves that it does not matter whether we choose M_1 or another marking from $[M_1]$. This follows from the definition of consistency and compatibility. □

When we construct an OE-graph by the algorithm in Prop. 2.5, we usually select those markings which we have found as successors during the execution of the for-loop. This implies that all the selected markings will be reachable.

In Sect. 3.1 we will introduce a special kind of equivalence specifications, derived from a concept called symmetry specifications. Such equivalence specifications are known to have stronger properties than ordinary equivalence specifications and hence we can optimise the algorithm in Prop. 2.5. This can be done because, under certain circumstances, we know in advance that two equivalent binding elements b_1 and b_2 will lead to equivalent markings. Hence there is no need to find the successor markings for both binding elements. We shall return to this optimisation in Sect. 3.5.

2.3 Proof Rules for OE-graphs

In this section we assume that we are dealing with a CP-net that has a consistent or compatible equivalence specification $(\approx_M, \approx_{BE})$ and a *finite* OE-graph OEG = (V, A, N). When this is the case, we can use the OE-graph and the corresponding SCC-graph to investigate the dynamic properties introduced in Chap. 4 of Vol. 1. This is done by modifying the proof rules from Sect. 1.4.

All our proof rules are valid both for consistent and for compatible equivalence specifications – unless explicitly stated otherwise. Some of the proofs in this section are a bit complicated and may be skipped by readers who are primarily interested in the practical application of CP-nets.

First we consider the reachability properties. For brevity we use M^c to denote the strongly connected component to which [M] belongs. This means we write M^c instead of $[M]^c$.

Proposition 2.6: For the **reachability properties** we have the following proof rules, valid for all $M_1,M_2 \in [[M_0\rangle]$:

(i) $[M_0\rangle \subseteq [V]$.
(ii) $M_2 \in [[M_1\rangle] \Leftrightarrow DPF([M_1],[M_2]) \neq \emptyset$.
(iii) $M_2 \in [[M_1\rangle] \Leftrightarrow DPF(M_1{}^c,M_2{}^c) \neq \emptyset$.
(iv) $M_2 \in [[M_1\rangle] \Leftarrow |SCC| = 1$.

Explanation: The four properties are similar to the properties in Prop. 1.12. However, it is now no longer possible to determine whether a marking M_2 is *itself* reachable from M_0 or from another marking M_1. By introducing the equivalence relation $\approx_M$ we have stated that we do not want to distinguish between the different members of the equivalence classes. The existence of a directed path with end node [M] tells us there exists a matching occurrence sequence that ends in a marking of [M] – but we do not know which marking.

Proof: The proof is a modification of the proof for Prop. 1.12. Property (i) is an immediate consequence of Def. 2.3 (i). Property (ii) follows from Prop. 2.4. Property (iii) follows from (ii) and from Prop. 1.11 (iv). Finally, property (iv) follows from (ii) and from Def. 1.8 (i). □

Above we have seen that we cannot use the OE-graph to determine the possible end nodes of occurrence sequences. We can only determine the equivalence classes of the end nodes. It is important to notice that there are no similar problems for the start nodes. The reason is that the consistency and compatibility properties in Def. 2.2 guarantee that, for each occurrence sequence starting in a member of $[M_1]$, we can construct an occurrence sequence that starts in any specified member of $[M_1]$ and contains markings and binding elements equivalent to those of the original occurrence sequence.

Proposition 2.7: For the **boundedness properties** we have the following proof rules, valid for all $X \subseteq TE$, all $p \in PI$, and all functions $F \in [\mathbb{M} \to A]$ where $(A,\leq)$ is an arbitrary set with a linear ordering relation:

(i) $\text{Best Upper Bound}(X) \leq \max_{M \in [V]} |(M|X)|$.
(ii) $\text{Best Upper Integer Bound}(p) \leq \max_{M \in [V]} |M(p)|$.
(iii) $\text{Best Upper Multi-set Bound}(p) \leq \max_{M \in [V]} M(p)$.
(iv) $\text{Best UpperBound}(F) \leq \max_{M \in [V]} F(M)$.

By replacing max with min and $\leq$ with $\geq$, we get four rules for lower bounds.

Explanation: The properties are similar to the properties in Prop. 1.13. However, now we investigate all members of [V]. This means our investigation may include non-reachable markings, and hence we are no longer sure to find the best possible bounds. If [V] is infinite, it may happen that a maximum/minimum value does not exist. Then we cannot use the corresponding proof rule.

It is often the case that the equivalence relation $\approx_M$ matches the boundedness properties which we want to investigate, in the sense that $M_1 \approx_M M_2$ implies that $|(M_1|X)| = |(M_2|X)|$, $|M_1(p)| = |M_2(p)|$, $M_1(p) = M_2(p)$ or $F(M_1) = F(M_2)$. It is then straightforward to see that it is sufficient to investigate one marking from each equivalence class in V. Moreover, we can replace all $\leq$ and all $\geq$ by $=$. This means we can use the OE-graph to find the best possible bounds.

Proof: The proof follows from Prop. 2.6 (i) and from the various boundedness definitions in Sect. 4.2 of Vol. 1. □

Proposition 2.8: For the **home properties** we have the following proof rules, valid for all $X \subseteq [[M_0\rangle]$ and all $M \in [[M_0\rangle]$:

(i) $[X] \in HS \Leftrightarrow SCC_T \subseteq X^c$.
(ii) $[X] \in HS \Rightarrow |SCC_T| \leq |X|$.

(iii) $M \in HM \Rightarrow SCC_T = \{M^c\}$.
(iv) $HM \neq \emptyset \Rightarrow |SCC_T| = 1$.
(v) $M_0 \in HM \Rightarrow |SCC| = 1$.

Explanation: The properties are similar to the properties in Prop. 1.14. However, for properties (i) and (ii) it is now no longer sufficient to include all members of X in the home space. Instead we must include all members of [X]. The reason is the same as explained for Prop. 2.6.

It should also be noted that (iii)–(v) are no longer bi-implications. To see that we cannot allow bi-implications, consider Fig. 1.2, which is an OE-graph for the resource allocation system. The OE-graph has only one strongly connected component, but M_0 is not a home marking due to the cycle counters.

Proof: **Property (i):** The proof is a modification of the proof for Prop. 1.14 (i). Assume that $SCC_T \subseteq X^c$ and let $M' \in [M_0\rangle$ be a reachable marking. We shall then prove that $[X] \cap [M'\rangle \neq \emptyset$ (see Def. 4.8 (ii) of Vol. 1).

From Prop. 1.11 (iii) we know there exists a terminal component $c \in SCC_T$ such that $DPF(M'^c,c) \neq \emptyset$. From our assumption we know that c contains an equivalence class with a marking $M \in X$. We thus have $DPF(M'^c,M^c) \neq \emptyset$ and by Prop. 2.6 (iii) we conclude that there exists a marking $M^* \in [M] \subseteq [X]$ such that $M^* \in [M'\rangle$. Hence $M^* \in [X] \cap [M'\rangle$.

Next assume that there exists a terminal component $c \in SCC_T$ with no elements from [X]. Let $C \in c$ be a node in this component and choose $M \in C$ such that M is reachable. This can be done due to Def. 2.3 (i). It is then easy to see that M is a reachable marking from which it is impossible to reach a marking in [X] (because it is impossible to find a directed path that leaves the terminal component c).

Property (ii): Assume that $[X] \in HS$. We then have:

$$|SCC_T| \leq |X^c| \leq |X|$$

where the first inequality follows from $SCC_T \subseteq X^c$ and the second follows from the definition of X^c.

Properties (iii)–(v): These properties follow from (i) and (ii) and the fact that $M \in HM$ implies $[M] \in HS$. □

We now introduce some notation allowing us to inspect how the different binding elements appear in OE-graphs and the corresponding SCC-graphs. The notation is analogous to that introduced for O-graphs. For a node $C_1 \in V$ we use $BE(C_1)$ to denote the set of all binding elements which appear in an output arc of C_1:

$$BE(C_1) = \{b \in BE \mid \exists C_2 \in V: (C_1, [b], C_2) \in A\}.$$

For a directed path $dp \in DP$ we use BE(dp) to denote the set of all binding elements that appear in an arc of dp:

$$BE(dp) = \{b \in BE \mid \exists C_1, C_2 \in dp: (C_1, [b], C_2) \in dp\}.$$

For a strongly connected component $c \in SCC$ we use BE(c) to denote the set of all binding elements that appear in an arc which starts in c:

$$BE(c) = \{b \in BE \mid \exists C_1 \in c\ \exists C_2 \in V: (C_1, [b], C_2) \in A\}.$$

Proposition 2.9: For the **liveness properties** we have the following proof rules, valid for all $M \in [M_0\rangle]$, all $X \subseteq BE$, and all $t \in T \cup TI$:

Consistent equivalence specification:

(i) M is dead $\Leftrightarrow$ [M] is terminal.
(ii) M is dead $\Leftrightarrow$ M^c is terminal and trivial.

(iii) [X] is dead in M $\Leftrightarrow$ $\forall c \in SCC$: $[DPF(M^c, c) = \emptyset \vee BE(c) \cap X = \emptyset]$.
(iv) [X] is live $\Leftrightarrow$ $\forall c \in SCC_T$: $BE(c) \cap X \neq \emptyset$.
(v) $[BE(t)] = BE(t)$ $\Rightarrow$ (t is live $\Leftrightarrow$ $\forall c \in SCC_T$: $t \in BE(c)$).

Compatible equivalence specification:

(vi) M is dead $\Rightarrow$ [M] is terminal.
(vii) M is dead $\Rightarrow$ M^c is terminal and trivial.

(viii) [X] is dead in M $\Rightarrow$ $\forall c \in SCC$: $[DPF(M^c, c) = \emptyset \vee BE(c) \cap X = \emptyset]$.
(ix) [X] is live $\Leftarrow$ $\forall c \in SCC_T$: $BE(c) \cap X \neq \emptyset$.

Explanation: Properties (i)–(v) are satisfied for a consistent equivalence specification. They are similar to properties (i)–(v) in Prop. 1.15. One difference is that properties (iii) and (iv) involve [X] instead of X. The reason is similar to that explained for Prop. 2.6. The existence of a directed path containing a class of binding elements [b] tells us there is an occurrence sequence that contains a binding element from [b] – but we do not know which binding element. Another difference is that (v) contains an extra assumption $[BE(t)] = BE(t)$. This means

(v) only can be used for equivalence specifications defined in such a way that two equivalent binding elements are guaranteed to belong to the same transition/transition instance.

Properties (vi)–(ix) are satisfied for a compatible equivalence specification. They are similar to (i)–(iv), but weaker. We now only have four implications instead of four bi-implications. The reason is that the OE-graph gives us no information about τ steps.

Proof: Let us first assume that the equivalence specification is consistent. We then have to prove four bi-implications.

Property (i): Straightforward consequence of Prop. 2.6 (ii).

Property (ii): The right-hand sides of (i) and (ii) are equivalent; see the proof of Prop. 1.15 (ii).

Property (iii): The proof is a modification of the proof of Prop. 1.15 (iii). Assume there exists a strongly connected component $c \in SCC$ that does not fulfil the requirement in the right-hand side of (iii). From $BE(c) \cap X \neq \emptyset$ we know there exists a node $C \in c$ which has a binding element $b \in X$ in an output arc. According to Def. 2.3 (ii) and Def. 2.2 (i) this means each marking $M' \in C$ has an enabled binding element from $[b] \subseteq [X]$. From $DPF(M^c,c) \neq \emptyset$ and Prop. 1.12 (iii) we know there exists a marking $M'' \in C$ such that $M'' \in [M\rangle$. Hence we have shown that [X] is non-dead in M.

Next assume that [X] is non-dead in M. This means there exists a finite occurrence sequence that starts in M and ends in a marking M' in which a binding element from [X] is enabled. From Prop. 2.6 (iii) we know that $DPF(M^c,M'^c) \neq \emptyset$ and from Def. 2.3 (ii) and Def. 2.2 (i) we know that $BE(M'^c) \cap X \neq \emptyset$. Hence we have shown that M'^c does not satisfy the right-hand side of (iii).

Property (iv): The proof is a modification of the proof of Prop. 1.15 (iv). Assume that $BE(c) \cap X \neq \emptyset$ for all terminal strongly connected components and let M be a reachable marking. We shall prove that [X] is non-dead in M (see Def. 4.10 (iii) of Vol. 1). From Prop. 1.11 (iii) we conclude that there exists a terminal strongly connected component c such that $DPF(M^c,c) \neq \emptyset$. This means the following is false:

$$DPF(M^c,c) = \emptyset \vee BE(c) \cap X = \emptyset$$

and thus we conclude from (iii) that [X] cannot be dead in M.

Next assume there exists a terminal strongly connected component c such that $BE(c) \cap X = \emptyset$. From (iii) it is easy to prove that [X] is dead in all markings M belonging to a node in c. Hence we conclude that X cannot be live.

Property (v): Direct consequence of property (iv) and the definition of liveness for a transition/transition instance.

Properties (vi)–(ix): A thorough examination tells us that the proofs of $\Rightarrow$ in (i)–(iii) and the proof of $\Leftarrow$ in (iv) only use arguments and propositions that are valid for both kinds of equivalence specifications when we replace Def. 2.2 (i) by Def. 2.2 (ii). □

Finally, let us consider the fairness properties. For a set of binding elements $X \subseteq BE$ we construct a subgraph of the OE-graph by deleting all arcs containing an equivalence class [b] where $b \in X$. We use $SCC_{OE\backslash X}$ to denote the set of all strongly connected components of this subgraph. Moreover, we use the notation defined above Prop. 2.9.

Proposition 2.10: For the **fairness properties** we have the following proof rules, valid for all $X \subseteq BE$:

Consistent equivalence specification:

(i) [X] is impartial $\Leftrightarrow \forall dc \in DCS: [BE(dc) \cap X \neq \emptyset]$.
(ii) [X] is fair $\Leftrightarrow \forall dc \in DCS: [BE(dc) \cap X \neq \emptyset \vee \forall C \in dc: BE(C) \cap X = \emptyset]$.
(iii) [X] is just $\Leftrightarrow \forall dc \in DCS: [BE(dc) \cap X \neq \emptyset \vee \exists C \in dc: BE(C) \cap X = \emptyset]$.

(iv) [X] is impartial $\Leftrightarrow \forall c \in SCC_{OE\backslash X}$: [c is trivial].
(v) [X] is fair $\Leftrightarrow \forall c \in SCC_{OE\backslash X}$: [c is trivial $\vee \forall C \in c: BE(C) \cap X = \emptyset$].
(vi) [X] is just $\Leftrightarrow \forall c \in SCC_{OE\backslash X}$: [c is trivial $\vee$
$\forall dc \in DCS(c)\ \exists C \in dc: BE(C) \cap X = \emptyset$].

Compatible equivalence specification:

No proof rules.

Explanation: The properties are similar to the properties in Prop. 1.16. The main difference is that we now use [X] instead of X. The reason is as explained for Prop. 2.9.

For a compatible equivalence specification we cannot use the OE-graph to investigate fairness properties. The reason is that the OE-graph gives us no information about τ steps.

Proof: **Property (i):** The proof is a modification of the proof of Prop. 1.16 (i). Assume that each simple directed cycle contains an element from X, and let $\sigma \in OSI$ be an infinite occurrence sequence (which starts in a reachable marking). We shall prove that $OC_{[X]}(\sigma) = \infty$ (see Def. 4.12 of Vol. 1). From Prop. 2.4 (i) we know there exists an infinite directed path $dp \in DPI$ that matches the infinite occurrence sequence σ. Since dp is infinite it must contain one of the directed cycles $dc \in DCS$ infinitely many times. From $BE(dc) \cap X \neq \emptyset$ we then conclude that $OC_{[X]}(\sigma) = \infty$.

Next assume that [X] is impartial and let $dc \in DCS$ be a simple directed cycle. We shall prove that dc contains an element from X. To do this we construct an infinite occurrence sequence $\sigma \in OSI$ which matches the cycle repeated infinitely often. We choose the start marking of σ so that it is reachable. This can be done due to Prop. 2.4 (ii). From the definition of impartiality, we know that $OC_{[X]}(\sigma) = \infty$. Hence we conclude that $BE(dc) \cap X \neq \emptyset$.

Properties (ii)–(iii): The proofs of (ii) and (iii) are rather straightforward modifications of the proof of (i). Hence they are omitted.

Properties (iv)–(vi): The OE-graph contains a simple directed cycle without elements from [X] iff $SCC_{OE\setminus X}$ has a non-trivial component. With this observation it is easy to verify that (iv)–(vi) follow from (i)–(iii), respectively. □

2.4 Example of OE-graphs: Distributed Data Base

In this section we study the OE-graph for the data base system from Sect. 1.3 of Vol. 1. For this system it is rather easy to calculate the number of nodes and arcs in the OE-graph, and thus we can study how the size of the OE-graph grows when we increase the number of data base managers.

First we have to define an equivalence specification, i.e., two equivalence relations $\approx_M$ and $\approx_{BE}$. The idea behind $\approx_M$ is as follows. We consider two markings M and M* to be equivalent iff there exists a bijection $\phi \in$ [DBM→DBM] such that we can obtain M* from M by replacing *all* appearances of c with ϕ(c) for all c∈DBM. The replacement is done in all DBM multi-sets and all MES multi-sets of the marking M. Intuitively, this means two markings are considered to be equivalent iff a bijective renaming of the data base managers maps one of the markings into the other. As an example, consider the two nodes (3,2,–) and (1,2,–) in Fig. 1.4. These two markings are equivalent, because the latter can be obtained from the former by means of the bijection which maps S(3) into S(1), S(2) into S(2), and S(1) into S(3).

With the above definition, it is straightforward to prove that $\approx_M$ is an equivalence relation. The relation is transitive because the functional composition $\phi_1 \circ \phi_2$ of two bijections is a bijection. The relation is symmetric because each bijection ϕ has an inverse function ϕ^{-1} which also is a bijection. Finally, the relation is reflexive because the identity function is a bijection.

The idea behind $\approx_{BE}$ is analogous. We consider two binding elements b and b* to be equivalent iff there exists a bijective renaming (of the elements of DBM) such that b is mapped into b*. From the CP-net it can be seen that this definition implies that two binding elements are equivalent iff they involve the same transition.

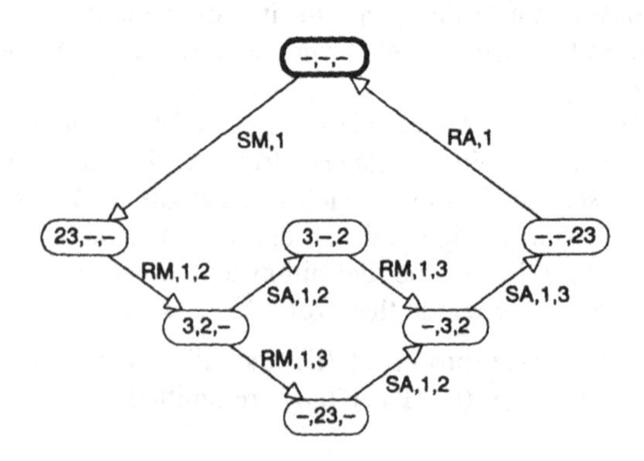

Fig. 2.1. OE-graph for data base system with 3 managers

It can easily be verified that ($\approx_M$,$\approx_{BE}$) is a consistent equivalence specification. The proof is similar to the consistency proof in Sect. 2.1. Hence we omit it. Now we are ready to construct the OE-graph for the data base system. The OE-graph for three data base managers looks as shown in Fig. 2.1. We inscribe each node [M] with one of the markings which belong to [M] (and we use the same encoding as in Fig. 1.4). Analogously, we inscribe each arc ($[M_1]$,[b],$[M_2]$) with one of the binding elements which belong to [b] (and again we use the encoding from Fig. 1.4).

The reader is encouraged to spend some time making a thorough comparison of the OE-graph in Fig. 2.1 and the O-graph in Fig. 1.4.

Figure 2.2 shows the OE-graph for the data base system with four different managers (and the same equivalence specification). It should be noted that the OE-graph in Fig. 2.2 is isomorphic to the occurrence graph shown in Fig. 5.2 of Vol. 1. The main difference between the two graphs is the fact that Fig. 2.2 uses a more condensed encoding of the markings.

From the OE-graph in Figs. 2.1 or 2.2 it is possible to verify most of the dynamic properties postulated for the data base system in Chap. 4 of Vol. 1 (see Exercise 2.2).

Now let us calculate the size of the OE-graph as a function of the number of data base managers n (we assume that $n \geq 3$). To do this we observe that, when we ignore the initial marking, the nodes form one half of a square, with n nodes in each direction. It is thus easy to calculate that we have the following number of nodes and arcs:

$$N_{OE}(n) = 1 + (1+2+3+\ldots+n) = 1 + n*(n+1)/2 = O(n^2).$$

$$A_{OE}(n) = 2 + 2*(1+2+3+\ldots+(n-1)) = 2 + (n-1)*n = O(n^2).$$

The results are shown for different numbers of data base managers in Fig. 2.3. It should be obvious that the OE-graphs are much smaller than the O-graphs.

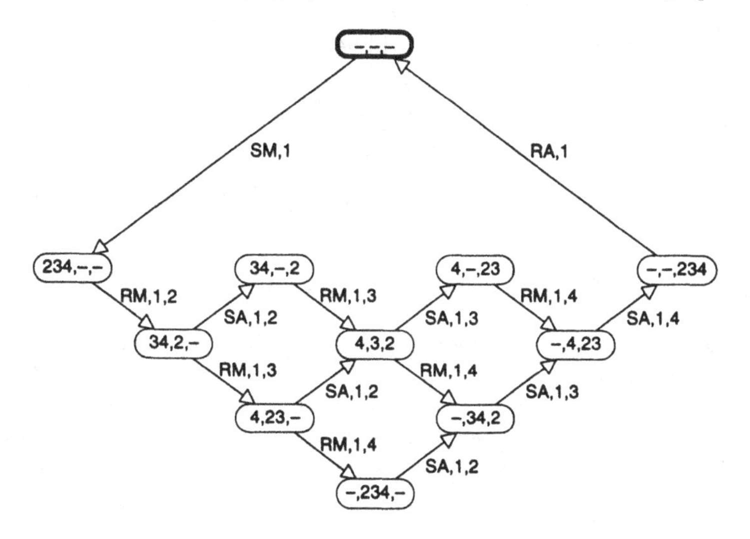

Fig. 2.2. OE-graph for data base system with 4 managers

Above, we have investigated the sizes of the O-graphs and the OE-graphs, i.e., the **space complexity** of the O-graph and OE-graph algorithms. However, it is equally important how much time it takes to generate the two kinds of occurrence graphs. This means we also want to investigate the **time complexity** of the two algorithms.

For both algorithms we shall assume that there is an efficient way to test whether two markings are identical/equivalent to each other and that the use of key functions (described in Sect. 1.7) means that each new marking only has to be compared to a small number of existing markings. With these assumptions it is easy to see that the time complexity is proportional to the number of times the body of the for-loop is executed. This means the time complexity of the O-graph algorithm is of order $O(n^2 * 3^n)$ because the for-loop is executed as many times as there are arcs. The time complexity of the OE-graph algorithm is of order $O(n^3)$ because the number of nodes is of order $O(n^2)$ while each of these (on average) has a number of enabled binding elements which is of order $O(n)$. The latter can be seen by an argument similar to that used to calculate the number of arcs in the O-graph (see Sect. 1.5).

	O-graph		OE-graph	
|DBM|	$N_O(n)$	$A_O(n)$	$N_{OE}(n)$	$A_{OE}(n)$
$O(n)$	$O(n*3^n)$	$O(n^2*3^n)$	$O(n^2)$	$O(n^2)$
2	7	8	4	4
3	28	42	7	8
4	109	224	11	14
5	406	1,090	16	22
6	1,459	4,872	22	32
7	5,104	20,426	29	44
8	17,497	81,664	37	58
9	59,050	314,946	46	74
10	196,831	1,181,000	56	92
15	71,744,536	669,615,690	121	212
20	23,245,229,341	294,439,571,680	211	382

Fig. 2.3. The size of the O-graphs and OE-graphs for the data base system

2.5 Example of OE-graphs: Producer/Consumer System

In the previous section we investigated an OE-graph constructed by means of a consistent equivalence specification. In the present section we investigate an OE-graph constructed from an equivalence specification that is only compatible. To do this we consider a CP-net which is similar to the producer/consumer system in Exercise 1.3 of Vol. 1 – including the modifications in questions (a) and (b).

Figure 2.4 shows the corresponding CP-net. The leftmost part of the net describes a set of producers, PROD. Each producer is able to produce and send a sequence of packages. To produce a package a producer first *Divides* the task into two subtasks, *Produce1* and *Produce2*, which are then performed concurrently. When the two subtasks have finished, the total package is *Produced* by combining the results of the two subtasks. We thus get a package (p,c,d1,d2) where p and c identify the producer and the intended consumer, while d1 and d2 are the data parts produced by *Produce1* and *Produce2*, respectively. We do not model how the two subtasks create the data values d1 and d2. This could be done, e.g., by reading the contents of a local data base. Instead we simply assume that there is a set of possible data values, DATA.

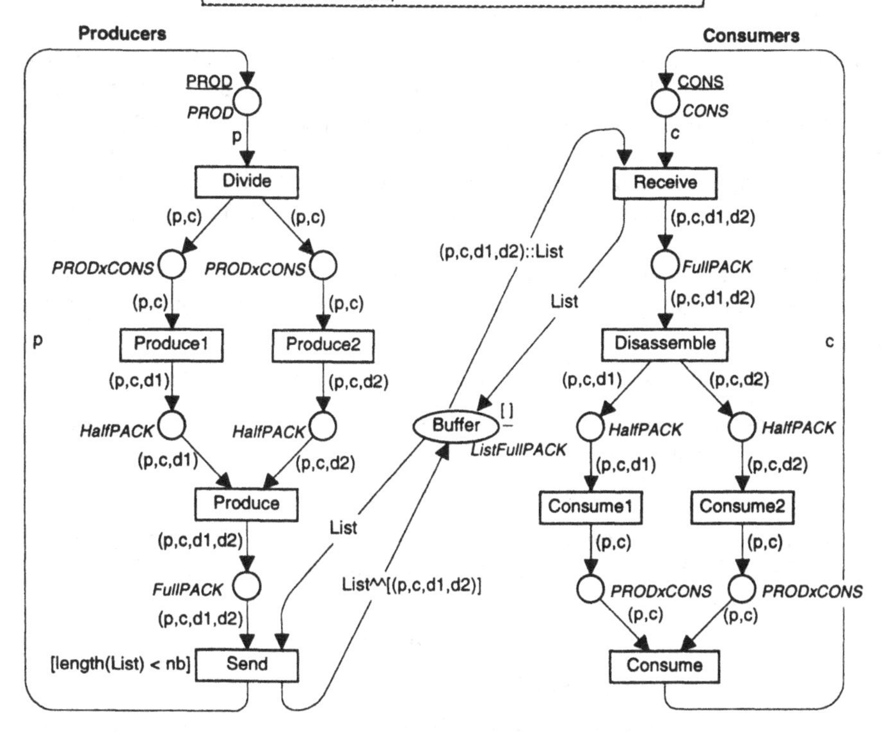

Fig. 2.4. CP-net describing a producer/consumer system

After the production, the producer *Sends* each package to the *Buffer*, where it is appended to the rear end of a list. The packages to be *Received* are taken from the front end of the list, and hence the *Buffer* works as a FIFO queue (first in, first out). The *Buffer* is bounded, since the guard of *Send* guarantees that the list will have a maximal length of nb packages.

The rightmost part of the net describes a set of consumers, CONS. The consumer part is analogous to the producer part. Each *Received* package is *Disassembled* into two parts, which are then concurrently consumed by two subtasks, *Consume1* and *Consume2*. When the two subtasks have finished, the entire package has been *Consumed*, and the consumer is ready to *Receive* a new package. We do not model how the two subtasks use the data values d1 and d2. They could, e.g., be used to update a local data base.

The cardinality of PROD, CONS, and DATA and the maximal size of the buffer are determined by four constants, np, nc, nd, and nb. We now calculate the number of nodes in the O-graph, as a function of these four constants. As we shall see, this is fairly straightforward. To enhance readability, we use p, c, d, and b, instead of np, nc, nd, and nb.

For each producer $p \in PROD$, there is the following number of possibilities for the position and colour of p-tokens (i.e., tokens where the first element of the tuple is p):

$$1 + c + 2*c*d + 2*c*d^2.$$

The first addend corresponds to the set of markings in which there is a p-token on the input place of *Divide*, while the second addend corresponds to the markings that have a p-token on each of the two output places. For the third addend there is a p-token on an input place of *Produce1* or *Produce2*, and a p-token on the output place of the other transition. Finally, the last addend corresponds to the markings where there is either a p-token on each of the two input places of *Produce* or a p-token on the output place.

Since the individual producer processes are totally independent of each other, we get the following number of possibilities for the markings of the producer part:

$$N_P(p,c,d) = (1 + c + 2*c*d + 2*c*d^2)^p = O((2*c*d^2)^p).$$

By a similar argument we get the following number of possibilities for the markings of the consumer part:

$$N_C(p,c,d) = (1 + p + 2*p*d + 2*p*d^2)^c = O((2*p*d^2)^c).$$

Finally, we have the following number of possibilities for the markings of the buffer (the i'th addend corresponds to the markings in which the buffer list contains $i-1$ elements):

$$\begin{aligned} N_B(p,c,d,b) &= 1 + p*c*d^2 + (p*c*d^2)^2 + \ldots + (p*c*d^2)^b \\ &= O((p*c*d^2)^b). \end{aligned}$$

The marking of the three parts of the net can be chosen independently of each other. To show this, we first find an occurrence sequence σ_1 leading from the initial marking to the desired marking of the consumer part. It is easy to see that we can choose σ_1 so that the buffer becomes empty. Next we append an occurrence sequence σ_2 that leads to the desired marking of the *Buffer* – without changing the marking of the consumer part. It is easy to see that we can choose σ_2 in such a way that all producers are returned to their initial state. Finally, we append an occurrence sequence σ_3 that leads to the desired marking of the producer part – without changing the marking of the consumer part and the *Buffer*. From the discussion above we conclude that the O-graph has the following number of nodes:

$$\begin{aligned} N_O(p,c,d,b) &= N_P(p,c,d) * N_C(p,c,d) * N_B(p,c,d,b) \\ &= O((2 * c * d^2)^p) * O((2 * p * d^2)^c) * O((p * c * d^2)^b) \\ &= O(2^{p+c} * p^{c+b} * c^{p+b} * d^{2*(p+c+b)}). \end{aligned}$$

The number of nodes in the O-graph grows very fast when p, c, d, and b grow. In Fig. 2.5 this is shown in the column labelled *O-graph*.

It should be obvious that we can analyse the producer/consumer system without considering the data values. The reason is that our model does not use the data values for any kind of synchronisation between processes. Hence the data values can be removed from the model without changing the behavioural properties of the CP-net. The situation is totally analogous to the removal of the cycle counters in the resource allocation system. However, this time there are two different ways in which the data values can be removed without changing the net inscriptions. One possibility is to set the constant nd = 1. Another possibility is to use the OE-graph formalism with an equivalence specification, where two markings/binding elements are defined to be equivalent iff they are identical when the data values are ignored. The two methods yield occurrence graphs of identical sizes. Using the first method we get O-graphs, while by the second method we get OE-graphs based on a consistent equivalence specification. The number of nodes in the graphs is determined by the following formula, where N_P, N_C, and N_B are the same as in the calculation of N_O:

p	c	b	d	O-graph	Consistent OE-graph	Compatible OE-graph
2	2	1	2	$9.03 * 10^6$	$7.32 * 10^4$	$4.05 * 10^2$
2	2	2	2	$1.45 * 10^8$	$3.07 * 10^5$	$1.70 * 10^3$
3	3	2	3	$1.28 * 10^{15}$	$1.53 * 10^9$	$3.73 * 10^5$
3	3	3	3	$1.04 * 10^{17}$	$1.38 * 10^{10}$	$3.36 * 10^6$
4	4	2	4	$3.61 * 10^{22}$	$1.03 * 10^{13}$	$1.07 * 10^8$
4	4	4	4	$2.37 * 10^{27}$	$2.64 * 10^{15}$	$2.73 * 10^{10}$
5	5	2	5	$2.82 * 10^{30}$	$9.19 * 10^{16}$	$3.94 * 10^{10}$
5	5	5	5	$6.88 * 10^{38}$	$1.44 * 10^{21}$	$6.15 * 10^{14}$
10	10	2	10	$7.80 * 10^{74}$	$1.43 * 10^{38}$	$6.80 * 10^{24}$
10	10	5	10	$7.80 * 10^{86}$	$1.43 * 10^{44}$	$6.80 * 10^{30}$

Fig. 2.5. The size of O-graphs and OE-graphs for the producer/consumer system

$$\begin{aligned} N_{CON}(p,c,b) &= N_P(p,c,1) * N_C(p,c,1) * N_B(p,c,1,b) \\ &= (1 + 5 * c)^p * (1 + 5 * p)^c \\ &\qquad * (1 + p * c + (p * c)^2 + \ldots + (p * c)^b) \\ &= O((5 * c)^p) * O((5 * p)^c) * O((p * c)^b) \\ &= O(5^{p+c} * p^{c+b} * c^{p+b}). \end{aligned}$$

In Fig. 2.5 this is shown in the column labelled *Consistent OE-graph.*

Although we have reduced the number of nodes, it is still very large even for small p, c, and b. Hence we want to obtain additional reductions. This can be done if, during the occurrence graph analysis, we are prepared to ignore some of the details of the production and consumption. This could make good sense if our primary goal is to validate the synchronisation/communication between producers and consumers via the bounded buffer. Then we may consider the occurrence of *Divide, Produce1, Produce2, Disassemble, Consume1* and *Consume2* to be of little interest, because they deal with the details of the production and consumption of data and not with the communication between the processes. Hence we make the occurrence of these transitions non-observable. This is done by defining a compatible equivalence specification ($\approx_M$, $\approx_{BE}$).

As $\approx_M$ and $\approx_{BE}$ we use the smallest equivalence relations that ignore the data values and guarantee that the occurrence of any of the six transitions mentioned above, in an arbitrary marking M_1, leads to a marking M_2 that is equivalent to M_1. With these definitions it is quite straightforward to prove that ($\approx_M$, $\approx_{BE}$) constitutes a compatible equivalence specification (see Exercise 2.7). Hence we can construct the corresponding OE-graph. From the discussion above and the

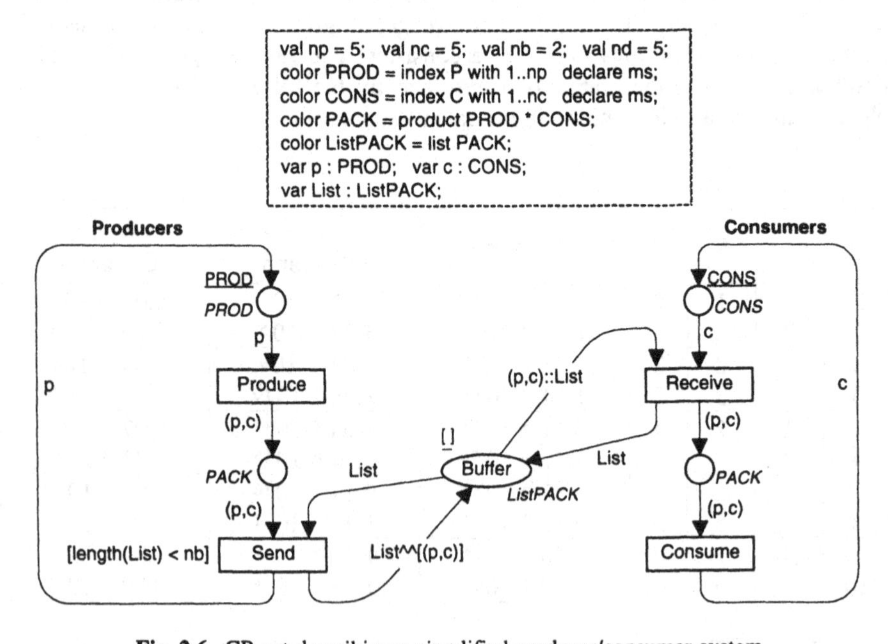

Fig. 2.6. CP-net describing a simplified producer/consumer system

definition of the OE-graph of a compatible equivalence specification, it can be seen that the OE-graph is isomorphic to the O-graph of the CP-net shown in Fig. 2.6.

Using the same kind of arguments as we used to calculate N_O and N_{CON}, we can easily calculate the number of nodes in the new OE-graph constructed from our compatible equivalence specification:

$$\begin{aligned} N_{COM}(p,c,b) &= N'_P(p,c) * N'_C(p,c) * N'_B(p,c,b) \\ &= (1+c)^p * (1+p)^c * (1 + p*c + (p*c)^2 + \ldots + (p*c)^b) \\ &= O(c^p) * O(p^c) * O((p*c)^b) \\ &= O(p^{c+b} * c^{p+b}). \end{aligned}$$

In Fig. 2.5 this is shown in the column labelled *Compatible OE-graph.*

In this section we have seen how a compatible equivalence specification can be used to obtain a reduced OE-graph that is considerably smaller than the OE-graph which we could obtain by a consistent equivalence specification. As with the resource allocation system, it can be argued that the same reduction could have been obtained by simplifying the CP-net before the start of the occurrence graph analysis (replacing the CP-net in Fig. 2.4 by the one in Fig. 2.6). This is correct for the producer/consumer example. However, OE-graphs based on compatible equivalence specifications can be used in a lot of situations where it is not obvious how to simplify the original CP-net.

For the producer/consumer system we can obtain an additional reduction by modifying the equivalence specification to allow bijective renamings of the elements in PROD and CONS in a similar way as we did for the DBM colour set in the data base system. Unfortunately, it is now no longer easy to make an exact formula for the number of nodes in the OE-graph. However, there are $p!*c!$ different renamings, and it can be proved that most markings by these renamings will be mapped into different markings (as long as the length of the buffer is comparable to the number of producers/consumers). Hence we can expect a reduction factor which is close to $p!*c!$.

Our definition of OE-graphs is deliberately made extremely general. The only property which we have required of the equivalence specification is consistency/compatibility. This leaves room for experiments with many different kinds of equivalence specifications. A few possibilities have been illustrated by the examples in this chapter. However, until now we have only rather limited practical experiences with the OE-graph method. So there is still a lot of practical work to be done, such as experimenting with different kinds of equivalence specifications.

All the examples presented in Chaps. 1–3 are small and too simple to be typical for practical CP-net applications. We have chosen examples that are easy to present and for which it is relatively easy to make analytic calculations of the sizes of the corresponding occurrence graphs. Even for our small examples, we have seen that the occurrence graphs become very big. Hence the reader may wonder whether it is possible at all to use the occurrence graph method for CP-nets with the complexity encountered in typical industrial applications. The size of an occurrence graph depends on the size of the CP-net, on the size of the colour sets involved and on the degree to which processes may interleave their executions. By experience, the last two factors seem to be much more important than the first one, and there exist several examples of CP-nets with a fairly large net structure for which we can construct occurrence graphs without any problems. Finally, it should be remembered that we can construct occurrence graphs for selected parts of a CP-net, such as one or more pages. This will often be a fast and easy way to debug new parts of a large model.

2.6 Labelled OE-graphs

An OE-graph does not contain any information about individual markings or binding elements. It only contains information about equivalence classes of markings and equivalence classes of binding elements. Hence it is impossible, e.g., to use the OE-graph to tell whether a given binding element b is enabled in a given marking M. The only thing we can tell is whether there exists some binding element in [b] which is enabled in some marking of [M].

Actually, the algorithm in Prop. 2.5 calculates a lot of information about individual markings and binding elements. For each marking M_1, selected from [Waiting], the algorithm constructs Next(M_1). This means the algorithm calculates all enabled binding elements and the corresponding successor markings. However, this information is not recorded in the OE-graph. It is used to calculate the output nodes and output arcs of [M_1]. Then the information is thrown away.

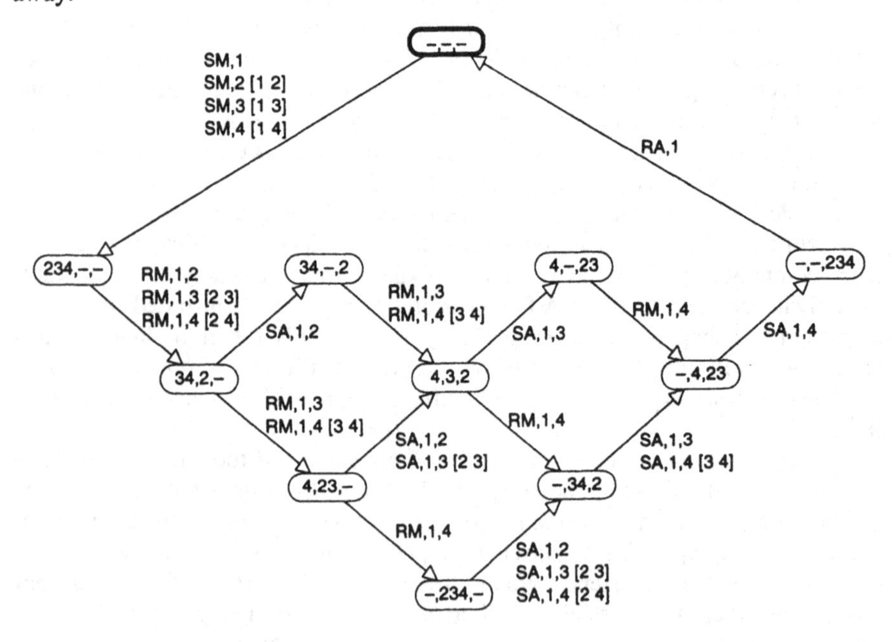

Fig. 2.7. Labelled OE-graph for data base system with 4 managers

As we have seen above, information about individual markings and binding elements may be useful in some situations. This is, in particular, true when we have an equivalence specification such that, for a marking $M_1^* \in [M_1]$, we can deduce information about Next(M_1^*) from Next(M_1). In Chap. 3 we consider a kind of OE-graph for which this is the case.

We define labelled OE-graphs below. Each such graph is identical to the corresponding OE-graph except that it retains the information about individual markings and binding elements.

A labelled OE-graph has two labelling functions M and L. The first of these functions maps each node $v \in V$ into a **node label** denoted by M_V. Analogously, the second function maps each arc $a \in A$ into an **arc label** denoted by L_a. By using M_V and L_a, instead of M(v) and L(a), we save space and also improve readability. Furthermore, it becomes easier to distinguish the labelling function M from a marking M. As in Vol. 1 we use X_S to denote the set of all subsets of a set X.

Definition 2.11: Let a CP-net and a consistent equivalence specification $(\approx_M, \approx_{BE})$ be given. A **labelled OE-graph** is a tuple OELG = (OEG, M, L) where OEG = (V, A, N) is the OE-graph, while $M \in [V \to \mathbb{M}]$ and $L \in [A \to (BE \times \mathbb{M})_S]$ are **labelling** functions satisfying the following properties:

(i) $\forall v \in V$: $M_V \in v$.

(ii) $\forall a \in A$: $L_a = \text{Next}(M_{s(a)}) \cap (BE \times d(a))$.

For a compatible equivalence specification we replace Next by Next_T.

A given CP-net and equivalence specification may have many different labelled OE-graphs (since we can choose the marking labels in different ways). It is straightforward to modify the algorithm in Prop. 2.5, so that it constructs a labelled OE-graph instead of an ordinary OE-graph. The modification can be done without increasing the time complexity of the algorithm.

In a labelled OE-graph we can deduce the equivalence class [M] of a given node from the node label. Analogously, we can deduce the three equivalence classes (C_1,B,C_2) of a given arc from the arc label and the node label of the source node. Hence the drawing of a labelled OE-graph needs no other text inscriptions than the node and arc labels. A labelled OE-graph of the data base system with four managers is shown in Fig. 2.7. It should be compared with the OE-graph in Fig. 2.2.

Each arc label has a line for each element $(b,M) \in \text{Next}(M_{s(a)}) \cap (BE \times d(a))$. The binding element b is represented in the same way as in Figs. 2.1 and 2.2, while the marking M is represented by specifying a permutation which map the marking in the label of the destination node into M. As an example, consider the leftmost label. The three lines specify that there are three enabled binding elements: *RM 1,2*, *RM 1,3*, and *RM 1,4*. The first binding element leads to the marking (34,2,–), which is identical to the label of the destination node. Hence there is no need for a permutation. The second binding element leads to the marking (24,3,–), which can be obtained from (34,2,–) by means of the permutation [2 3] mapping d_2 into d_3, d_3 into d_2, and all other managers into themselves. Analogously, the third binding element leads to the marking (23,4,–), which can be obtained from (34,2,–) by means of the permutation [2 4] mapping d_2 into d_4, d_4 into d_2, and all other managers into themselves.

2.7 Computer Tools for OE-graphs

This section describes how the construction and analysis of OE-graphs are supported by computer tools. In Sect. 1.7 we discussed tool support for O-graphs. It should be obvious that most of that discussion is also valid for tools dealing with OE-graphs. In the present section we shall not repeat the arguments of Sect. 1.7. Instead we focus on the additional requirements needed to support OE-graphs.

Analysis of OE-graphs

The analysis of OE-graphs can be done in exactly the same way as the analysis of O-graphs, i.e., by means of SearchNodes, SearchArcs, and SearchComponents. This is the case because the right-hand sides of the proof rules in Sect. 2.3 are similar to the right-hand sides of the proof rules of Sect. 1.4 (except for small notational differences). For the boundedness properties this is only true when we choose the equivalence specification so as to match the boundedness properties we want to investigate (see the explanation of Prop. 2.7). Otherwise, we need a way to calculate all the members of the equivalence class to which a marking belongs.

Construction of OE-graphs

The construction of OE-graphs requires a method to test whether a given pair of markings (or binding elements) are equivalent to each other or not. A very straightforward solution is to leave the implementation of the equivalence tests to the user. This means the user has to write two SML functions, **EquivMark** and **EquivBE**. The first function takes two markings and tells us whether they are equivalent or not. The second function does the same, but for two binding elements. To write the two SML functions, the user applies a set of predefined functions that allow him to access the individual parts of the markings (or binding elements).

With the above approach it is straightforward to modify an O-graph tool to become an OE-graph tool – as can be seen by comparing the two algorithms in Props. 1.4 and 2.5. For a consistent equivalence specification the tool processes a node [M] by calculating the set Next(M). This is done in exactly the same way as for an O-graph. However, when a new node/arc is to be inserted in the OE-graph, it is tested whether there is already a node/arc which is equivalent to it. These tests are performed by means of EquivMark and EquivBE.

For a compatible equivalence specification the situation is similar. The only difference is that now we have to calculate $\text{Next}_\tau(M)$ instead of Next(M). This means we have to find not only the markings that are directly reachable from M, but all the markings that are reachable via an occurrence sequence where all but the last binding element are non-observable. This can be done by constructing a small occurrence graph, where we start in M and only find successors of the nodes belonging to the same equivalence class as M. In theory this may yield an infinite or very large subgraph of the O-graph, but in practice this will probably rarely happen.

When we become more experienced with OE-graphs, it will also become appropriate to make a more direct tool support of some of the more common and useful kinds of equivalence specifications. This can be done by allowing the user to define the equivalence specification by means of a high-level description, from which the OE-graph tool can automatically construct EquivMark and EquivBE. Another possibility is to supply a library of predefined functions that makes it easier for the user to write EquivMark and EquivBE (for certain kinds of equivalence specifications).

Key functions

In Sect. 1.7 it was explained how the occurrence graph tool represents markings by means of a set of search trees. To get an efficient OE-graph construction it is necessary that the keys in the search trees are constructed in such a way that two equivalent markings are guaranteed to have identical keys. Otherwise, we will have to compare each new marking with all existing marking records (and not only those which have the same key). For similar reasons, we want equivalent binding elements to have identical keys.

The three standard key functions described in Sect. 1.7 will work for many equivalence specifications. For the resource allocation system we can use any of the three key functions. For the data base system we can use K_1 or K_2. We could also use K_3, but this is only because all coefficients are either 0 or 1. Otherwise, a permutation of the colours might change the order in which the coefficients appear in the list. For the data base system there is no difference between the information contents of K_1, K_2, and K_3. Hence we can just as well choose K_1, which is the simplest.

For the compatible equivalence specification of the producer/consumer system we can use the standard key functions for the multi-set search trees, while we have to modify the key functions used for page records and marking records.

Soundness of OE-graphs

Now let us consider to what extent an OE-graph tool will be able to check whether a proposed equivalence specification is sound, i.e., consistent or compatible with the CP-net. In general this is not possible, because it requires a mathematical proof in which the consistency/compatibility property of Def. 2.2 has to be checked for *all* pairs of markings in *all* possible equivalence classes. This can only be machine supported if we have an extremely general and powerful theorem prover. However, there are many situations where we can make a full or partial check of the soundness. We now discuss some of these situations.

First let us assume that each reachable marking has a finite number of enabled binding elements, and hence a finite number of direct successor markings, reachable by the occurrence of a single binding element. This assumption is satisfied by most CP-nets used in practice. In this case it is easy to implement an automatic check telling whether two concrete markings M_1 and M_2 (given by the user) are consistent with each other, i.e., satisfy the property in Def. 2.2 (i). We first use EquivMark to test whether M_1 and M_2 are equivalent to each other. If they are non-equivalent there is nothing more to be done. Otherwise, we calculate Next(M_1) and Next(M_2), which are finite due to our assumption. The calculation is done in exactly the same way as the tool finds the direct successor markings of a node, during the creation of the occurrence graph. Finally, we compare Next(M_1) and Next(M_2). This is done by means of EquivMark and EquivBE. For a compatible equivalence specification the situation is similar. The only difference is that now we use Def. 2.2 (ii) and hence we have to calculate $\text{Next}_\tau(M_1)$ and $\text{Next}_\tau(M_2)$. We have already seen how this can be done.

Above, we have discussed how the OE-graph tool can help the user to check the consistency/compatibility of two given markings. This can never prove the soundness of an entire equivalence specification, of course, but in practice it will be a good help in finding possible mistakes.

As discussed above, later versions of the OE-graph tool may make a more thorough support of selected kinds of common and useful equivalence specifications. For some of these kinds, it will be possible to make a full check of their consistency/compatibility, because this can be determined from structural properties of the CP-net. As a trivial example, let us consider an equivalence specification where two markings are equivalent iff they are identical when a given set of places are ignored, while two binding elements are equivalent iff they are identical. When the ignored places have no output arcs, the equivalence specification is known to be consistent.

In the next chapter we discuss equivalence specifications, which are derived from permutations of the individual colour sets. Also, in this case it will be possible for many CP-nets to check the consistency of an entire equivalence specification via an automatic check, based on local structural properties. We will return to this in Sect. 3.5.

Bibliographical Remarks

To the best of our knowledge, all the material in Chap. 2 is original, and has not yet been published elsewhere. There are several other authors who work with different methods to reduce occurrence graphs. However, none of these approaches are particularly close to the ideas behind occurrence graphs with equivalence classes – while some of them are close to the ideas behind occurrence graphs with symmetries. Hence we shall consider these alternative approaches in the bibliographical remarks of Chap. 3.

We believe that the basic ideas behind occurrence graphs with equivalence classes are directly applicable to occurrence graphs of arbitrary transition systems.

Exercises

Exercise 2.1.
Consider the resource allocation system from Fig. 1.1 and the OE-graph in Fig. 1.2.

(a) Check that the OE-graph is correct.

(b) Investigate whether you can use the OE-graph and the proof rules in Prop. 2.7 to verify the upper and lower bounds postulated at the end of Sect. 4.2 of Vol. 1.

(c) Investigate whether you can use the OE-graph and the proof rules in Prop. 2.8 to verify the home properties postulated at the end of Sect. 4.3 of Vol. 1.

(d) Investigate whether you can use the OE-graph and the proof rules in Prop. 2.9 to verify the liveness properties postulated at the end of Sect. 4.4 of Vol. 1.

(e) Investigate whether you can use the OE-graph and the proof rules in Prop. 2.10 to verify the fairness properties postulated at the end of Sect. 4.5 of Vol. 1.

Exercise 2.2.
Consider the data base system from Sect. 1.3 of Vol. 1 and the OE-graphs in Figs. 2.1 and 2.2.

(a) Check that the OE-graphs are correct. This can be done by comparing them to the O-graph in Fig. 1.4.

(b) Investigate whether you can use the OE-graphs and the proof rules in Prop. 2.7 to verify the upper and lower bounds postulated at the end of Sect. 4.2 of Vol. 1.

(c) Investigate whether you can use the OE-graphs and the proof rules in Prop. 2.8 to verify the home properties postulated at the end of Sect. 4.3 of Vol. 1.

(d) Investigate whether you can use the OE-graphs and the proof rules in Prop. 2.9 to verify that SM and RA are strictly live, while RM and SA are live.

(e) Investigate whether you can use the OE-graphs and the proof rules in Prop. 2.10 to verify the fairness properties postulated at the end of Sect. 4.5 of Vol. 1.

Exercise 2.3.
Consider the philosopher system from Sect. 1.6.

(a) Define a suitable equivalence specification and construct an OE-graph for the philosopher system with $|PH| = |CS| = 5$. What is the size of the OE-graph?

(b) Use the OE-graph and the proof rules in Prop. 2.7 to investigate the boundedness properties of the philosopher system.

(c) Use the OE-graph and the proof rules in Prop. 2.8 to investigate the home properties of the philosopher system.

(d) Use the OE-graph and the proof rules in Prop. 2.9 to investigate the liveness properties of the philosopher system.

(e) Use the OE-graph and the proof rules in Prop. 2.10 to investigate the fairness properties of the philosopher system.

(f) Repeat (a)–(e) with $|PH| = |CS| = 7$ and with $|PH| = |CS| = 9$ (if possible).

Exercise 2.4.
Consider the telephone system from Sect. 3.2 of Vol. 1. This exercise is only worth attempting if you have access to an occurrence graph tool. Even though the net is rather small, it will take too long to produce the OE-graphs if this has to be done manually.

(a) Define a suitable equivalence specification and construct an OE-graph for the telephone system with $|U| = 3$. What is the size of the OE-graph?

(b) Investigate whether you can use the OE-graph and the proof rules in Prop. 2.7 to verify the upper and lower bounds postulated at the end of Sect. 4.2 of Vol. 1.

(c) Investigate whether you can use the OE-graph and the proof rules in Prop. 2.8 to verify the home properties postulated at the end of Sect. 4.3 of Vol. 1.

(d) Investigate whether you can use the OE-graph and the proof rules in Prop. 2.9 to verify the liveness properties postulated at the end of Sect. 4.4 of Vol. 1. Investigate whether any of the transitions are strictly live.

(e) Investigate whether you can use the OE-graph and the proof rules in Prop. 2.10 to verify the fairness properties postulated at the end of Sect. 4.5 of Vol. 1.

(f) Repeat (a)–(e) with $|U| = 4$ and with $|U| = 5$ (if possible).

Exercise 2.5.
Consider the process control system from Exercise 4.6 of Vol. 1. This exercise is only worth attempting if you have access to an occurrence graph tool. Even though the net is rather small, it will take too long to produce the OE-graphs if this has to be done manually.

(a) Define a suitable equivalence specification and construct an OE-graph for the process control system with $|PROC| = |RES| = 2$. What is the size of the OE-graph?

(b) Use the OE-graph and the proof rules in Prop. 2.7 to investigate the boundedness properties of the process control system.

(c) Use the OE-graph and the proof rules in Prop. 2.8 to investigate the home properties of the process control system.

(d) Use the OE-graph and the proof rules in Prop. 2.9 to investigate the liveness properties of the process control system.

(e) Use the OE-graph and the proof rules in Prop. 2.10 to investigate the fairness properties of the process control system.

(f) Repeat (a)–(e) with $|PROC| = |RES| = 3$ and with $|PROC| = |RES| = 4$ (if possible).

Exercise 2.6.
Consider the modified ring network from Exercise 1.6. This exercise is only worth attempting if you have access to an occurrence graph tool. Even though the net is rather small, it will take too long to produce the OE-graphs if this has to be done manually.

(a) Define a suitable equivalence specification and construct an OE-graph for the ring network with NoOfBuffers = 2. What is the size of the OE-graph?

(b) Use the OE-graph and the proof rules in Prop. 2.7 to investigate the boundedness properties of the modified ring network.

(c) Use the OE-graph and the proof rules in Prop. 2.8 to investigate the home properties of the modified ring network.

(d) Use the OE-graph and the proof rules in Prop. 2.9 to investigate the liveness properties of the modified ring network.

(e) Use the OE-graph and the proof rules in Prop. 2.10 to investigate the fairness properties of the modified ring network.

(f) Repeat (a)–(e) with NoOfBuffers = 3 and with NoOfBuffers = 4 (if possible).

Exercise 2.7.
Consider the producer/consumer system from Fig. 2.4. This exercise is only worth attempting if you have access to an occurrence graph tool. Even though the net is rather small, it will take too long to produce the OE-graphs if this has to be done manually.

(a) Prove that the equivalence specification ($\approx_M$, $\approx_{BE}$) is compatible.

(b) Construct an OE-graph for the producer/consumer system with p = c = b = d = 2.

(c) Use the OE-graph and the proof rules in Prop. 2.7 to investigate the boundedness properties of the producer/consumer system.

(d) Use the OE-graph and the proof rules in Prop. 2.8 to investigate the home properties of the producer/consumer system.

(e) Use the OE-graph and the proof rules in Prop. 2.9 to investigate the liveness properties of the producer/consumer system.

(f) Use the OE-graph and the proof rules in Prop. 2.10 to investigate the fairness properties of the producer/consumer system.

Chapter 3

Occurrence Graphs with Symmetries

In the previous chapter we investigated how to obtain smaller occurrence graphs by means of equivalence specifications. In the present chapter we investigate a related concept called symmetry specifications. Each consistent symmetry specification determines a consistent equivalence specification, and thus we can talk about OE-graphs for consistent symmetry specifications. This kind of OE-graph is called an OS-graph. OS-graphs have stronger proof rules than the ordinary OE-graphs investigated in Chap. 2. The reason is that a consistent symmetry specification guarantees a number of behavioural properties that a consistent equivalence specification does not.

OS-graphs can be used to verify nearly all the dynamic properties defined in Vol. 1. The construction of OS-graphs and the associated verification of dynamic properties can be fully automated. Hence OS-graphs provide a very powerful and relatively easy-to-use method to analyse the properties of a given CP-net.

In a similar way as we introduced labelled OE-graphs, we introduce labelled OS-graphs. A labelled OS-graph contains exactly the same information as the corresponding O-graph. In fact it is possible to construct the O-graph from the labelled OS-graph.

Section 3.1 contains an informal introduction and the formal definition of symmetry specifications. The introduction is made by means of the data base and philosopher systems, i.e., two of the systems which we used to introduce O-graphs and OE-graphs. Section 3.2 contains a number of detailed proof rules, i.e., propositions that allow us to prove (or disprove) different CP-net properties by inspection of different OS-graph properties. The proof rules are modifications of the proof rules for OE-graphs. Section 3.3 discusses how symmetry specifications can be inherited from algebraic groups of permutations, defined for the individual colour sets. Section 3.4 extends OS-graphs by adding node and arc labels giving information about the behaviour of individual markings and binding elements. It is proved that each labelled OS-graph contains exactly the same information as the corresponding O-graph. Finally, Sect. 3.5 discusses how the construction and analysis of OS-graphs are supported by computer tools.

3.1 Symmetry Specifications

In Sect. 2.4 we considered an equivalence specification $(\approx_M, \approx_{BE})$ for the distributed data base system. We defined two markings to be equivalent iff a bijective renaming of the data base managers maps one of the markings into the other. Analogously, we defined two binding elements to be equivalent iff a bijective renaming of the data base managers maps one of the binding elements into the other.

As mentioned in Sect. 2.4, it is straightforward to prove that the above equivalence specification is consistent. Whenever we make a bijective renaming of the data base managers, we get a marking in which the same transition instances are enabled. The only thing different is that the bindings involved change in the same way as the marking. This means we can obtain the new enabled set of bindings by using the same bijective renaming as we used for the marking.

For a bijection $\phi \in [DBM \rightarrow DBM]$ and a marking M we use $\phi(M)$ to denote the equivalent marking obtained from M by the renaming specified by ϕ. Analogously, for a binding element b we use $\phi(b)$ to denote the equivalent binding element obtained from b by the bijective renaming specified by ϕ.

Inspired by the discussion above, we now define what we mean by a symmetry specification. We use $\circ$ to denote functional composition: $(g \circ f)(x) = g(f(x))$.

Definition 3.1: A **symmetry specification** for a CP-net is a set of functions $\Phi \subseteq [\mathbb{M} \cup BE \rightarrow \mathbb{M} \cup BE]$ such that:

(i) $(\Phi, \circ)$ is an algebraic group.
(ii) $\forall \phi \in \Phi$: $(\phi \mid \mathbb{M}) \in [\mathbb{M} \rightarrow \mathbb{M}] \wedge (\phi \mid BE) \in [BE \rightarrow BE]$.

Each element of Φ is called a **symmetry**.

We require $(\Phi, \circ)$ to be an algebraic group (with the identity function as neutral element). This means the following properties must be fulfilled for all ϕ, ϕ_1, ϕ_2, $\phi_3 \in \Phi$:

(a) $\phi_1 \circ \phi_2 \in \Phi$ (closed).
(b) $\exists \phi^{-1} \in \Phi$: $\phi \circ \phi^{-1} = \phi^{-1} \circ \phi = Id$ (inverse element).
(c) $\phi \circ Id = Id \circ \phi = \phi$ (neutral element).
(d) $\phi_1 \circ (\phi_2 \circ \phi_3) = (\phi_1 \circ \phi_2) \circ \phi_3$ (associative).

It is easy to see that the group properties imply that all ϕ are bijections, and so are all the restrictions $\phi \mid \mathbb{M}$ and $\phi \mid BE$.

Now let us again consider the data base system and the set of functions $\Phi \subseteq [\mathbb{M} \cup BE \rightarrow \mathbb{M} \cup BE]$ determined from bijections over DBM in the way described at the beginning of this section. It is easy to verify that Φ constitutes a symmetry specification. We only have to prove the group properties (a)–(d) and they follow from standard properties of bijections.

Intuitively, we consider each symmetry to represent a transformation rule by which we can modify a given marking (or a given binding element) such that it is mapped into a new marking (or a binding element) with similar properties to those of the old one. We then say that the two markings (or the two binding elements) are **symmetrical** to each other, because there exists a symmetry mapping one of them into the other. From the following proposition we see that being symmetrical is an equivalence relation.

Proposition 3.2: The relation $\approx_M \subseteq \mathbb{M} \times \mathbb{M}$ defined by:

(i) $M \approx_M M^* \Leftrightarrow \exists \phi \in \Phi: M = \phi(M^*)$

is an equivalence relation on the set of all markings $\mathbb{M}$.

The relation $\approx_{BE} \subseteq BE \times BE$ defined by:

(ii) $b \approx_{BE} b^* \Leftrightarrow \exists \phi \in \Phi: b = \phi(b^*)$

is an equivalence relation on the set of all binding elements BE.

Proof: The proof is a straightforward consequence of the group properties listed below Def. 3.1. Property (a) implies that the relations are transitive. Property (b) implies that the relations are symmetric. The existence of a neutral element in property (c) implies that the relations are reflexive. □

From the above definition it is obvious that we can test whether two markings or two binding elements are equivalent to each other by trying all symmetries. However, there are often much more efficient ways to determine the question of equivalence. We shall return to this in Sect. 3.5 when we discuss computer support for OS-graphs.

We expect two symmetrical markings (or binding elements) to have similar behavioural properties. This means we expect two symmetrical markings to have sets of enabled binding elements that are symmetrical to each other, and to have sets of directly reachable markings that are symmetrical to each other. This is formalised as follows:

Definition 3.3: A symmetry specification Φ is **consistent** iff the following properties are satisfied for all symmetries $\phi \in \Phi$, all markings $M_1, M_2 \in [M_0\rangle$ and all binding elements $b \in BE$:

(i) $\phi(M_0) = M_0$.

(ii) $M_1 [b\rangle M_2 \Leftrightarrow \phi(M_1) [\phi(b)\rangle \phi(M_2)$.

The first property tells us that each symmetry maps the initial marking into itself. This means $[M_0] = \{M_0\}$. The second property tells us that we for each step $M_1 [b\rangle M_2$ and each symmetry $\phi \in \Phi$ can find a symmetrical step $\phi(M_1) [\phi(b)\rangle \phi(M_2)$. It should be noted that the different elements of the new step may be totally or partially identical to the original ones, because ϕ may map M_1, b and/or M_2 into themselves.

Proposition 3.4: Each consistent symmetry specification Φ determines a consistent equivalence specification $(\approx_M, \approx_{BE})$.

Proof: According to Def. 2.2 we have to prove that $[\text{Next}(M_1)] = [\text{Next}(M_2)]$ for all markings M_1 and M_2 which are equivalent to each other. To prove $\subseteq$ it is sufficient to show that for each pair $(b,M) \in \text{Next}(M_1)$ we can find an equivalent pair $(b^*, M^*) \in \text{Next}(M_2)$. However, this follows directly from Def. 3.3 (ii). We simply choose $(b^*, M^*) = (\phi(b), \phi(M))$ where ϕ is a symmetry such that $M_2 = \phi(M_1)$. The proof in the other direction is similar. □

Definition 3.5: Let a CP-net and a consistent symmetry specification Φ be given. The **occurrence graph with symmetries**, also called the **OS-graph**, is the OE-graph obtained from the equivalence specification $(\approx_M, \approx_{BE})$ determined by Φ.

From Def. 3.5 it follows that OS-graphs constitute a particular case of OE-graphs. This means we can construct OS-graphs by the same algorithm as we construct OE-graphs, i.e., by means of the algorithm in Prop. 2.5.

It is easy to see that the symmetry specification for the data base system is consistent and determines an equivalence specification $(\approx_M, \approx_{BE})$ identical to the one we considered in Sect. 2.4. Hence we conclude that Figs. 2.1 and 2.2 are OS-graphs for the data base system.

Note that we require the modeller to specify the symmetry specification and to prove that it is consistent. In Sect. 3.5 we discuss how the latter can be supported by computer tools. An alternative approach would be to leave all the work to the occurrence graph tool, which then could try to obtain the "best possible" symmetry specification. Such an approach is indeed possible and is pursued by some researchers. We have chosen the first approach for two different reasons. Firstly, it is easier to implement, because it is simpler to check consistency of a given symmetry specification than it is to construct a symmetry specification from scratch. Secondly, our approach allows the detection of modelling errors, in a similar way to a compiler that detects type errors. For the data base system the modeller expects all data base managers to be interchangeable with each other without influencing the behaviour of the model. Hence it is straightforward for him to choose the symmetry specification that allows all permutations of DBM. If this choice makes the binding elements of one of the transitions inconsistent, the modeller wants to be warned, because it indicates that he may have made a modelling error. Using the alternative approach the modeller would not be warned. Instead the tool would construct a symmetry specification that would be different from the one the modeller had in mind.

Next, we consider the philosopher system introduced in Sect. 1.6. It has two colour sets, PH and CS, both of size n. As for the data base system, we can use bijective renamings of the colour sets to define a symmetry specification. For the data base system we allowed all possible permutations of DBM. For the philosopher we only allow rotations. For each $k \in 1..n$ we use R_k to denote the renaming which maps each philosopher ph_i and each chopstick cs_i into $ph_{i \oplus k}$ and $cs_{i \oplus k}$, respectively. $\oplus$ denotes "cyclic addition", i.e., $i \oplus k = ((i+k-1) \bmod n)+1$.

It is easy to verify that the rotations $\Phi_R = \{R_k \mid k \in 1..n\}$ determine a set of functions $\Phi \subseteq [\mathbb{M} \cup BE \rightarrow \mathbb{M} \cup BE]$ satisfying the properties in Def. 3.1. Hence we have a symmetry specification. It is also straightforward to show that Φ satisfies the properties in Def. 3.3. Hence it is consistent.

Each philosopher has a special relationship to his two nearest neighbours, because he shares a chopstick with each of them. This is the reason why we only permit rotations. If we allowed arbitrary permutations we would consider a marking in which ph_1 and ph_3 are eating to be equivalent with one in which ph_1 and ph_2 are eating. This is obviously not what we want, and it is easy to see that it would violate the consistency property in Def. 3.3 (ii).

The OS-graph for five philosophers is shown in Fig. 3.1. Node #1 represents the initial marking, which constitutes an equivalence class of its own (due to Def. 3.3 (i)). Node #2 represents an equivalence class with five markings. We have inscribed node #2 by the marking in which ph_1 is eating. However, the node also represents the other four markings in which exactly one philosopher is eating. Also node #3 represents an equivalence class with five markings. We have inscribed node #3 by the marking in which ph_1 and ph_3 are eating. However, the node also represents the four other markings in which exactly two philosophers are eating. Each arc represents an equivalence class of binding elements. There are only two such equivalence classes – one containing the five binding elements of *Take Chopsticks* and one containing the five binding elements of *Put Down-Chopsticks*. By looking at the number of eating philosophers in the source and destination node of an arc, it is trivial to see whether the arc represents the first or the second equivalence class. Hence we have omitted all arc inscriptions.

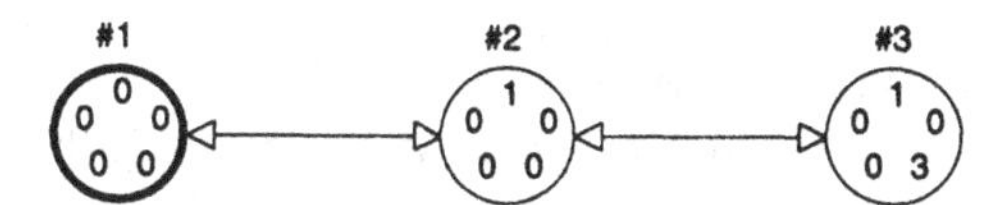

Fig. 3.1. OS-graph for five philosophers

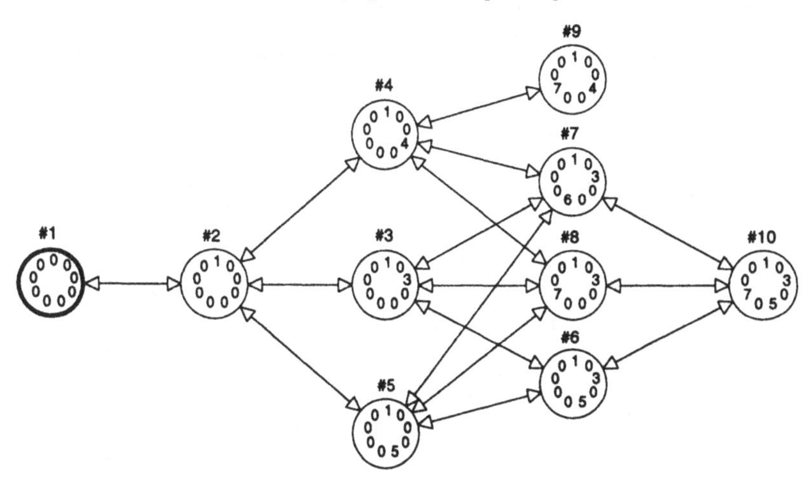

Fig. 3.2. OS-graph for nine philosophers

The OS-graph for nine philosophers is shown in Fig. 3.2. Node #1 represents an equivalence class with a single marking and node #9 represents an equivalence class with three markings. All other nodes represent an equivalence class with nine markings. Each arc represents an equivalence class with nine binding elements.

Figure 3.3 compares the size of the OS-graph with the size of the O-graph for different numbers of philosophers. It can be seen that the OS-graphs are significantly smaller than the O-graphs.

In the following section we shall see that OS-graphs have a set of proof rules that are more powerful than the standard proof rules introduced in Sect. 2.3. The reason is that a consistent symmetry specification guarantees some properties that a consistent equivalence specification does not. As an example, it is easy to see that $M_1 [b\rangle M_2$ implies that for a given marking $M_2^* \in [M_2]$ there exists a marking $M_1^* \in [M_1]$ and a binding element $b^* \in [b]$ such that $M_1^* [b^*\rangle M_2^*$. Analogously, it can be seen that $M_1 [b\rangle M_2$ implies that for a given binding element $b^* \in [b]$ there exists a marking $M_1^* \in [M_1]$ and a marking $M_2^* \in [M_2]$ such that $M_1^* [b^*\rangle M_2^*$. Neither of these properties are fulfilled for arbitrary consistent equivalence specifications.

When we apply a symmetry $\phi \in \Phi$ to all the markings and all the binding elements of a finite or infinite occurrence sequence, we still have an occurrence sequence:

Proposition 3.6: Let a consistent symmetry specification Φ be given. The following properties are satisfied for all $M, M_1, M_2, M_3, \ldots \in \mathbb{M}$, all $b_1, b_2, \ldots \in BE$ and all $\phi \in \Phi$:

(i) $M_1 [b_1\rangle M_2 [b_2\rangle M_3 \ldots M_n [b_n\rangle M_{n+1} \Leftrightarrow$
$\phi(M_1) [\phi(b_1)\rangle \phi(M_2) [\phi(b_2)\rangle \phi(M_3) \ldots \phi(M_n) [\phi(b_n)\rangle \phi(M_{n+1})$.

(ii) $M_1 [b_1\rangle M_2 [b_2\rangle M_3 \ldots \Leftrightarrow \phi(M_1) [\phi(b_1)\rangle \phi(M_2) [\phi(b_2)\rangle \phi(M_3) \ldots$

(iii) $M_2 \in [M_1\rangle \Leftrightarrow \phi(M_2) \in [\phi(M_1)\rangle$.

(iv) $M \in [M_0\rangle \Leftrightarrow \phi(M) \in [M_0\rangle$.

	O-graph		OS-graph	
$\lvert PH \rvert$	$N_O(n)$	$A_O(n)$	$N_{OS}(n)$	$A_{OS}(n)$
2	3	4	2	2
3	4	6	2	2
4	7	16	3	4
5	11	30	3	4
6	18	60	5	8
7	29	112	5	10
8	47	208	8	18
9	76	378	10	32

Fig. 3.3. The size of the O-graphs and OS-graphs for the philosopher system

Proof: Property (i) and (ii) follow by repeated use of Def. 3.3 (ii). Property (iii) is a direct consequence of (i). Property (iv) follows from (iii) and from Def. 3.3 (i). □

3.2 Proof Rules for OS-graphs

In this section we assume we are dealing with a CP-net that has a consistent symmetry specification Φ determining an equivalence specification with a *finite* OS-graph OSG = (V, A, N). As explained in Sect. 3.1, it is now possible to find a set of proof rules that are more powerful than the standard OE-graph proof rules presented in Sect. 2.3.

Below we introduce, explain, and prove the new set of proof rules. To enhance the overview, we also list those proof rules, which are identical to proof rules of OE-graphs. Some of the proofs in this section are a bit complicated and may be skipped by readers who are primarily interested in the practical application of CP-nets.

First we consider the reachability properties. For a directed path dp (of the OS-graph), we use Sym(dp) to denote the fact that dp contains at least one node [M] that is symmetrical, i.e., consists of only one marking. Analogously, we use Sym(c) to denote that a strongly connected component c contains at least one symmetrical node. Finally, we use Sym(dp) to denote that a directed path dp (of the SCC-graph) contains at least one component c satisfying Sym(c).

Proposition 3.7: For the **reachability properties** we have the following proof rules, valid for all $M_1, M_2 \in [M_0\rangle$:

(i) $[M_0\rangle = [V]$.

(ii) $M_2 \in [[M_1\rangle] \Leftrightarrow DPF([M_1],[M_2]) \neq \emptyset$.
(iii) $M_2 \in [[M_1\rangle] \Leftrightarrow DPF(M_1{}^c, M_2{}^c) \neq \emptyset$.

(iv) $M_2 \in [[M_1]\rangle \Leftrightarrow DPF([M_1],[M_2]) \neq \emptyset$.
(v) $M_2 \in [[M_1]\rangle \Leftrightarrow DPF(M_1{}^c, M_2{}^c) \neq \emptyset$.

(vi) $M_2 \in [M_1\rangle \Leftarrow \exists dp \in DPF([M_1],[M_2])$: Sym(dp).
(vii) $M_2 \in [M_1\rangle \Leftarrow \exists dp \in DPF(M_1{}^c, M_2{}^c)$: Sym(dp).

(viii) $M_2 \in [M_1\rangle \Leftarrow |SCC| = 1$.

Explanation: The first property tells us that a marking is reachable iff it belongs to [V]. This is a stronger property than Prop. 2.6 (i).

Properties (ii)–(iii) are identical to Prop. 2.6 (ii)–(iii). They allow us to investigate whether, for two given markings M_1 and M_2, there exist one or more finite occurrence sequences starting in M_1 and ending in some member of $[M_2]$. Analogously, properties (iv)–(v) deal with occurrence sequences starting in some member of $[M_1]$ and ending in M_2. We use $[[M_1]\rangle$ to denote the set of all the markings reachable from a member of $[M_1]$. Finally, properties (vi)–(viii) deal with occurrence sequences starting in M_1 and ending in M_2.

Proof: **Property (i):** From Prop. 2.6 (i) we know that $[M_0\rangle \subseteq [V]$. Hence it is sufficient to prove that $[V] \subseteq [M_0\rangle$. To do this, let us assume that $M \in [V]$. This means there exists an equivalence class $C \in V$ such that $M \in C$. From Def. 2.3 (i) we know that $C \cap [M_0\rangle \neq \emptyset$, and thus there exists a marking M^* such that $M \approx_M M^*$ and $M^* \in [M_0\rangle$. From $M \approx_M M^*$ and Def. 3.2 (i) we know there exists a symmetry such that $M = \phi(M^*)$. From $M^* \in [M_0\rangle$ and Prop. 3.6 (iv) we know that $M = \phi(M^*) \in [M_0\rangle$.

Properties (ii)–(iii): Identical to Prop. 2.6 (ii)–(iii).

Properties (iv)–(v): From (ii) and (iii) it follows that it is sufficient to prove that $[[M_1\rangle] = [[M_1]\rangle$.

Let us first assume that $M_2 \in [[M_1\rangle]$, i.e., that there exists a marking $M_2^* \in [M_2]$ such that $M_2^* \in [M_1\rangle$. From $M_2^* \in [M_2]$ we know there exists a symmetry $\phi \in \Phi$ such that $M_2 = \phi(M_2^*)$. From $M_2^* \in [M_1\rangle$ and Prop. 3.6 (iii) we know that $M_2 = \phi(M_2^*) \in [\phi(M_1)\rangle$. Hence we have found a marking $M_1^* = \phi(M_1) \in [M_1]$ such that $M_2 \in [M_1^*\rangle$ and from this we conclude that $M_2 \in [[M_1]\rangle$.

The proof in the other direction is analogous. This time we choose $\phi \in \Phi$ such that $M_1 = \phi(M_1^*)$.

Property (vi): Assume that the right-hand side of (vi) is satisfied. From Prop. 2.4 (ii) we know that dp has a matching occurrence sequence:

$$M_1 = M_1^* [b_1^*\rangle M_2^* [b_2^*\rangle M_3^* \ldots M_n^* [b_n^*\rangle M_{n+1}^* \in [M_2]$$

where each of the markings M_i^* belongs to the equivalence class of the i'th node in the path dp. The occurrence sequence can be split into two parts:

$$M_1 [\sigma_1\rangle M_k^* [\sigma_2\rangle M_{n+1}^* \in [M_2]$$

where M_k^* is a symmetrical marking. From $M_{n+1}^* \in [M_2]$ we know there exists a symmetry $\phi \in \Phi$ such that $\phi(M_{n+1}^*) = M_2$. By Prop. 3.6 (i) we get a finite occurrence sequence:

$$M_1 [\sigma_1\rangle M_k^* = \phi(M_k^*) [\phi(\sigma_2)\rangle \phi(M_{n+1}^*) = M_2.$$

Hence we conclude that $M_2 \in [M_1\rangle$.

Property (vii): From the definition of SCC-graphs it is easy to verify that the right-hand side of (vii) implies the right-hand side of (vi).

Property (viii): It is easy to verify that the right-hand side of (viii) implies the right-hand side of (vii). To do this, we use Def. 3.3 (i), which tells us that $[M_0] = \{M_0\}$. □

Proposition 3.8: For the **boundedness properties** we have the following proof rules, valid for all $X \subseteq TE$, all $p \in PI$, and all functions $F \in [\mathbb{M} \rightarrow A]$ where $(A, \leq)$ is an arbitrary set with a linear ordering relation:

(i) $\text{Best Upper Bound}(X) = \max_{M \in [V]} |(M|X)|$.
(ii) $\text{Best Upper Integer Bound}(p) = \max_{M \in [V]} |M(p)|$.
(iii) $\text{Best Upper Multi-set Bound}(p) = \max_{M \in [V]} M(p)$.
(iv) $\text{Best Upper Bound}(F) = \max_{M \in [V]} F(M)$.

By replacing max with min, we get four rules for lower bounds.

Explanation: The properties are similar to the properties in Prop. 2.7. The only difference is that now we are able to find the best possible bounds (because $[M_0\rangle = [V]$ instead of $[M_0\rangle \subseteq [V]$).

It is often the case that the equivalence relation $\approx_M$ matches the boundedness properties we want to investigate, in the sense that $M_1 \approx_M M_2$ implies that $|(M_1|X)| = |(M_2|X)|$, $|M_1(p)| = |M_2(p)|$, $M_1(p) = M_2(p)$ or $F(M_1) = F(M_2)$. It is then straightforward to see that it is sufficient to investigate one marking from each equivalence class in V.

Proof: The proof follows from Prop. 3.7 (i) and from the various boundedness definitions in Sect. 4.2 of Vol. 1. □

Proposition 3.9: For the **home properties** we have the following proof rules, valid for all $X \subseteq [[M_0\rangle]$ and all $M \in [[M_0\rangle]$:

(i) $[X] \in HS \Leftrightarrow SCC_T \subseteq X^c$.
(ii) $[X] \in HS \Rightarrow |SCC_T| \leq |X|$.

(iii) $M \in HM \Rightarrow SCC_T = \{M^c\}$.
(iv) $HM \neq \emptyset \Rightarrow |SCC_T| = 1$.

(v) $X \in HS \Leftarrow SCC_T \subseteq X^c \wedge \forall c \in SCC_T\text{: } Sym(c)$.
(vi) $M \in HM \Leftarrow SCC_T = \{M^c\} \wedge Sym(M^c)$.

(vii) $M_0 \in HM \Leftrightarrow |SCC| = 1$.

Explanation: The first four properties are identical to Prop. 2.8 (i)–(iv). Properties (v)–(vi) give us sufficient conditions for home spaces and home markings. They require each terminal strongly connected component to contain an equivalence class with only one element and an equivalence class with a marking from the home space/home marking. Finally, property (vii) deals with the particular case in which there is only one strongly connected component.

Proof: **Properties (i)–(iv):** Identical to Prop. 2.8 (i)–(iv).

Property (v): The proof is a modification of the proof for Prop. 2.8 (i). Assume that the right-hand side of (v) is satisfied and let $M' \in [M_0\rangle$ be a reachable marking. We shall then prove that $X \cap [M'\rangle \neq \emptyset$ (see Def. 4.8 (ii) of Vol. 1).

From Prop. 1.11 (iii) we know there exists a terminal component $c \in SCC_T$ such that $DPF(M'^c, c) \neq \emptyset$. From our assumption we know that c contains an equivalence class with a marking $M \in X$. This means $DPF(M'^c, M^c) \neq \emptyset$ and $Sym(M^c)$. By Prop. 3.7 (vii) we then conclude that $M \in [M'\rangle$. Hence $M \in X \cap [M'\rangle$.

Property (vi): Immediate consequence of (v).

Property (vii): To prove $\Leftarrow$ we show that M_0 satisfies the right-hand side of (vi). This follows from Def. 3.3 (i). The other direction is identical to Prop. 2.8 (v). □

Proposition 3.10: For the **liveness properties** we have the following proof rules, valid for all $M \in [M_0\rangle$, all $X \subseteq BE$, and all $t \in T \cup TI$:

(i) M is dead $\Leftrightarrow$ [M] is terminal.
(ii) M is dead $\Leftrightarrow$ M^c is terminal and trivial.

(iii) [X] is dead in M $\Leftrightarrow$ $\forall c \in SCC$: $[DPF(M^c, c) = \emptyset \vee BE(c) \cap X = \emptyset]$.
(iv) [X] is live $\Leftrightarrow$ $\forall c \in SCC_T$: $BE(c) \cap X \neq \emptyset$.
(v) $[BE(t)] = BE(t) \Rightarrow$ (t is live $\Leftrightarrow$ $\forall c \in SCC_T$: $t \in BE(c)$).

(vi) X is live $\Leftarrow$ $\forall c \in SCC_T$: $[BE(c) \cap X \neq \emptyset \wedge Sym(c)]$.
(vii) $|SCC| = 1 \Rightarrow$ (X is live $\Leftrightarrow$ $BE(M_0{}^c) \cap X \neq \emptyset$).

Explanation: The first five properties are identical to Prop. 2.9 (i)–(v). Property (vi) gives us a sufficient condition for liveness. It requires each terminal strongly connected component to contain an equivalence class with only one element and an equivalence class where a binding element from X appears in one of the output arcs. Finally, property (vii) tells us that when there is only one strongly connected component, a set of binding elements is live iff an element from it appears in one of the arcs.

Proof: **Properties (i)–(v):** Identical to Prop. 2.9 (i)–(v).

Property (vi): Assume the right-hand side of (vi) is satisfied and let M be a reachable marking. We shall prove that X is non-dead in M (see Def. 4.10 (iii) of Vol. 1).

From Prop. 1.11 (iii) we know there exists a terminal strongly connected component $c \in SCC_T$ such that $DPF(M^c, c) \neq \emptyset$. From $BE(c) \cap X \neq \emptyset$ we know that c has an arc (C_1, B, C_2) such that $B \subseteq [X]$, which by Def. 2.3 (ii) implies the existence of a marking $M_1 \in C_1$ and a binding element $b \in [X]$ such that $M_1 [b\rangle$. We can find a symmetry $\phi \in \Phi$ such that $\phi(b) \in X$. From Prop. 3.7 (vii) we know that $\phi(M_1)$ is reachable from M and from Def. 3.3 (ii) we know that $\phi(M_1) [\phi(b)\rangle$. Hence we conclude that X cannot be dead in M.

Property (vii): It is easy to verify that $|SCC| = 1$ and $BE(M_0{}^c) \cap X \neq \emptyset$ implies the right-hand side of (vi). Hence we have shown $\Leftarrow$ (in the bi-implication).

The other direction is also easy. $BE(M_0{}^c) \cap X = \emptyset$ implies that no element from [X] appears in the OS-graph. Hence X is dead in M_0. □

For the fairness properties we do not get more powerful proof rules. A consistent symmetry specification has the same proof rules as a consistent equivalence specification; see Prop. 2.10.

3.3 Permutation Symmetries

The definition of a symmetry specification in Sect. 3.1 is deliberately made extremely general. Our only requirement is the consistency properties in Def. 3.3. This leaves room for experiments with many different kinds of symmetry specifications. In the present section we investigate a very important class of symmetries, called **permutation symmetries**. These symmetries can be obtained from bijective renamings, i.e., permutations of colours, in a similar way as we did for the data base system in Sect. 2.4 and the philosopher system in Sect. 3.1. We will show that each consistent specification of permutation symmetries determines a consistent specification of ordinary symmetries. Hence we will be able to use all the results of Sect. 3.2 for permutation symmetries.

The basic ideas behind permutation symmetries are simple. For each colour set $S \in \Sigma$ we define a set of allowable permutations, SG(S), which is required to be a subgroup of the set of all permutations of S (using functional composition as the composition law). The subgroup SG(S) is called the **symmetry group** of S and each permutation in SG(S) is called a **colour symmetry** of S. From the colour symmetries of the individual colour sets we will derive symmetries for markings and binding elements (details will be explained below). By requiring each set of colour symmetries to be a subgroup, we guarantee that the derived symmetries (of markings and binding elements) will form an algebraic group. This means Def. 3.1 will be satisfied, and hence we have a symmetry specification.

It will often be the case that we allow all possible permutations of a given colour set. This was done for the DBM colour set of the data base system. However, in some situations it may be necessary to make a more restrictive definition of the set of colour symmetries in order to satisfy the consistency property of Def. 3.3 (ii). This was done for the PH and CS colour sets of the philosopher system where we only allowed rotations. There may also be colour sets for which we do not want to allow any renaming at all. This means the only colour symmetry we allow is the identity function.

We have now introduced the three most commonly used examples of symmetry groups, allowing all **permutations**, all **rotations**, and nothing but the identity function. The latter means that all colours are **fixed**. However, there are other more complex kinds of symmetry groups. As an example, we can obtain a smaller OS-graph for the philosopher system if, in addition to rotations, we also allow all permutations that amount to a flip (mirror) operation. We shall return to this in Sect. 3.4.

When we specify a permutation symmetry, we shall distinguish between atomic colour sets and structured colour sets. An **atomic colour set** is a colour set defined without reference to other existing colour sets. In contrast to this, a

structured colour set is defined, from a set of **base colour sets,** by means of a structuring mechanism, e.g., cartesian product, record, union, list, or subset. We use Σ_{AT} to denote the atomic colour sets while we use Σ_{ST} to denote the structured colour sets. For the data base system in Sect. 2.4 there are two atomic colours sets, DBM and E, and there are also two structured colour sets, PR and MES.

For a structured colour set we do not define the colour symmetries explicitly. Instead we derive the colour symmetries of the structured colour set from the colour symmetries of the base colour sets. For the data base system we derive the colour symmetries of MES from those of PR, which we again derive from those of DBM. More precisely, this means each DBM colour symmetry $\phi_{DBM} \in$ SG(DBM) determines a MES colour symmetry ϕ_{MES} defined by:

$$\phi_{MES}(s,r) = (\phi_{DBM}(s),\phi_{DBM}(r))$$

for all $(s,r) \in$ MES.

Having outlined the basic ideas behind a specification of permutation symmetries, we now give the formal definition:

Definition 3.11: A **permutation symmetry specification** is a function SG that maps each atomic colour set $S \in \Sigma_{AT}$ into a subgroup SG(S) of the set of all permutations of S.

A **permutation symmetry** for SG is a function ϕ that maps each atomic colour set $S \in \Sigma_{AT}$ into a permutation $\phi_S \in SG(S)$.

SG(S) is called the **symmetry group** of S while ϕ_S is called a **colour symmetry**. The set of all permutation symmetries for SG is denoted by Φ_{SG}.

As explained above, we derive colour symmetries for the structured colour sets from those of the base colour sets. This means we extend the domain of each permutation symmetry $\phi \in \Phi_{SG}$ from Σ_{AT} to Σ. In Def. 3.12 we show how to handle the structuring mechanisms of CPN ML, i.e., products, records, unions, lists and subsets (see Sect. 1.4 of Vol. 1). If another language is used for the declarations and inscriptions of CP-nets, the details will be slightly different, but the basic ideas will remain the same.

It should be noted that a base colour set may itself be structured, and hence it may be necessary to use Def. 3.12 in a recursive way. For example, consider the data base system where the colour symmetries of the subset colour set MES are derived from the colour symmetries of the product colour set PROD, which again are derived from the colour symmetries of the atomic colour set DBM.

Definition 3.12: For each permutation symmetry $\phi \in \Phi_{SG}$ and each structured colour set $S \in \Sigma_{ST}$ we define ϕ_S as follows:

(i) Product $S = \text{product } A_1 * A_2 * \ldots * A_n$:

$$\phi_S((a_1, a_2, \ldots, a_n)) = (\phi_{A_1}(a_1), \phi_{A_2}(a_2), \ldots, \phi_{A_n}(a_n)).$$

(ii) Record $S = \text{record } s_1{:}A_1 * s_2{:}A_2 * \ldots * s_n{:}A_n$:

$$\phi_S(\{s_1{:}a_1, s_2{:}a_2, \ldots, s_n{:}a_n\}) = \{s_1{:}\phi_{A_1}(a_1), s_2{:}\phi_{A_2}(a_2), \ldots, s_n{:}\phi_{A_n}(a_n)\}.$$

(iii) Union $S = \text{union } s_1{:}A_1 + s_2{:}A_2 + \ldots + s_n{:}A_n + c_1 + c_2 + \ldots + c_m$:

$$\phi_S(a) = \begin{cases} s_i(\phi_{A_i}(a_i)) & \text{iff } a = s_i(a_i) \quad \text{where } i \in 1..n \\ a & \text{iff } a = c_i \quad \text{where } i \in 1..m. \end{cases}$$

(iv) List $S = \text{list } A$:

$$\phi_S([a_1, a_2, \ldots, a_n]) = [\phi_A(a_1), \phi_A(a_2), \ldots, \phi_A(a_n)].$$

(v) Subset $S = \text{subset } A \text{ with } [a_1, a_2, \ldots, a_n]$ or $S = \text{subset } A \text{ by } F$:

$$\phi_S(a) = \phi_A(a)$$

and we require $\phi_A(S) \subseteq S$.

Above, we have considered colour symmetries, i.e., functions which map colours into colours. To obtain a symmetry – as defined in Def. 3.1 – it is necessary to obtain functions which map markings into markings and binding elements into binding elements. To do this we introduce linear functions between multi-sets. This is a general concept that is useful for many different purposes, e.g., place invariants (see Sect. 5.2 of Vol. 1).

Definition 3.13: Let S and R be two sets. A function $F \in [S_{MS} \to R_{MS}]$ is **linear** iff it satisfies the following property for all $m_1, m_2 \in S_{MS}$:

$$F(m_1 + m_2) = F(m_1) + F(m_2).$$

The set of all linear functions from S_{MS} to R_{MS} is denoted by $[S_{MS} \to R_{MS}]_L$.

It is easy to verify that a linear function also satisfies the following properties, for all $n \in \mathbb{N}$ and all $m, m_1, m_2 \in S_{MS}$:

(i) $F(n * m) = n * F(m)$.

(ii) $F(\emptyset) = \emptyset$.

(iii) $m_1 \leq m_2 \Rightarrow F(m_1) \leq F(m_2)$.

Proposition 3.14: Each function $F \in [S \to R]$ determines a unique linear function $F^* \in [S_{MS} \to R_{MS}]_L$ defined by:

$$F^*(m) = \sum_{s \in S} m(s)`F(s).$$

F* is called the **linear extension** of F.

Proof: The proof is a straightforward consequence of Def. 3.13 and Def. 2.2 (i) of Vol. 1. Hence it is omitted. □

Usually, we denote the extended function F^* by the same name as the original function F.

Next we show how a permutation symmetry can be used to derive functions for markings, bindings, and binding elements. $\phi_{C(p)}$ is the linear extension of the colour symmetry of C(p), while Type(v) is the type of the variable v (see Sect. 2.2 of Vol. 1).

Definition 3.15: Let a permutation symmetry $\phi \in \Phi_{SG}$, a marking $M \in \mathbb{M}$, a binding b, and a binding element $((t,i),b) \in BE$ be given. Then we define $\phi(M)$, $\phi(b)$ and $\phi((t,i),b)$ as follows:

(i) $\forall p \in PI$: $\phi(M)(p) = \phi_{C(p)}(M(p))$.
(ii) $\forall v \in Var(t)$: $\phi(b)(v) = \phi_{Type(v)}(b(v))$.
(iii) $\phi((t,i),b)) = ((t,i),\phi(b))$.

It is easy to see that ϕ maps each marking into a marking. However, it is not necessarily the case that ϕ maps each binding/binding element into a binding/binding element. The reason is that $G(t)<\phi(b)>$ may be false and then $\phi(b)$ is not a binding – although it assigns a value of correct type to each variable (see Def. 2.6 of Vol. 1).

Definition 3.16: A permutation symmetry specification SG is **consistent** iff the following properties are satisfied for all $\phi \in \Phi_{SG}$, all $t \in T$ and all $a \in A$:

(i) $\phi(M_0) = M_0$.
(ii) $\forall b \in B(t)$: $\phi(b) \in B(t)$.
(iii) $\forall b \in B(t(a))$: $E(a)<\phi(b)> = \phi(E(a)<b>)$.

The first property tells us that each permutation symmetry maps the initial marking into itself. The second property tells us that each permutation symmetry maps binding elements into binding elements. Finally, the third property tells us that it does not matter whether we use a permutation symmetry before or after the evaluation of an arc expression. Intuitively, the last two properties require that guards and arc expressions treat symmetrical bindings in a "symmetrical way". We cannot have an asymmetrical guard such as $[p \neq ph_3]$ and we cannot have an asymmetrical arc expression such as: if $p = ph_3$ then 1`p else empty.

It is important to notice that the properties in Def. 3.16 are structural properties. This means they can be verified without considering all the possible occurrence sequences. This is in contrast to Def. 3.3 (ii) which must be verified for all reachable markings.

We are now ready to prove the main result of this section. It tells us that each permutation symmetry specification (see Def. 3.11) that is consistent (see Def. 3.16) by means of Def. 3.15 determines a symmetry specification (see Def. 3.1) that is consistent (see Def. 3.3). Hence we have achieved our goal: to be able to specify symmetry specifications by means of symmetry groups.

Theorem 3.17: When a permutation symmetry specification SG is consistent, it determines a consistent symmetry specification Φ in which each symmetry $\phi \in \Phi$ is determined from a permutation symmetry of Φ_{SG} via Def. 3.15.

Proof: We first prove that Φ is a symmetry specification, i.e., that it satisfies Def. 3.1. The group property in Def. 3.1 (i) follows from the subgroup property in Def. 3.11. The first part of Def. 3.1 (ii) follows from Def. 3.15 (i) and the fact that $\phi_{C(p)}$ maps from $C(p)_{MS}$ into $C(p)_{MS}$. Finally, the second part of Def. 3.1 (ii) follows from Defs. 3.15 (ii)–(iii) and 3.16 (ii).

Next we prove that the symmetry specification is consistent, i.e., that it satisfies Def. 3.3. Property 3.3 (i) is identical to property 3.16 (i). Hence we concentrate on property 3.3 (ii). It is sufficient to prove that $M_1 [b\rangle M_2$ implies $\phi(M_1) [\phi(b)\rangle \phi(M_2)$ for all $\phi \in \Phi$. Then the other direction follows by means of the inverse ϕ^{-1}.

We first prove that $M_1 [b\rangle$ implies $\phi(M_1) [\phi(b)\rangle$. Thus we assume that $M_1 [b\rangle$ for some $M_1 \in [M_0\rangle$ and some $b=(t',b') \in BE$. By Def. 3.6 of Vol. 1 we have:

$$\forall p'' \in PIG: \sum_{p' \in p''} E(p',t')\langle b'\rangle \leq M_1(p''),$$

which by property (iii) below Def. 3.13 implies that the following is satisfied for all $\phi \in \Phi$:

$$\forall p'' \in PIG: \phi_{C(p'')}\Big(\sum_{p' \in p''} E(p',t')\langle b'\rangle\Big) \leq \phi_{C(p'')}(M_1(p''))$$

where $C(p'')$ is the common colour set of the places which contribute to p". From the linearity of $\phi_{C(p'')}$ we then get:

$$\forall p'' \in PIG: \sum_{p' \in p''} \phi_{C(p'')}(E(p',t')\langle b'\rangle) \leq \phi_{C(p'')}(M_1(p'')),$$

which by Defs. 3.16 (iii) and 3.15 (i) is equivalent to:

$$\forall p'' \in PIG: \sum_{p' \in p''} E(p',t')\langle \phi(b')\rangle \leq \phi(M_1)(p'')$$

which by Def. 3.6 of Vol. 1 implies that $\phi(M_1) [\phi(b)\rangle$ as required.

Next we can prove that $M_1 [b\rangle M_2$ implies $\phi(M_1) [\phi(b)\rangle \phi(M_2)$. However, this is done in a way which is totally analogous to the proof above, and thus we shall omit it. □

Examples of permutation symmetries

We have introduced a framework above that allows the user to define symmetry specifications by defining a symmetry group for each atomic colour set. The method can be used for many of the CP-nets considered in this book.

For the data base system we allow all permutations of DBM while E is fixed. For the telephone system we allow all permutations of U. For the producer/consumer system we allow all permutations of PROD and all permutations of CONS, and we either remove DATA before the analysis or allow all permutations of it.

For the philosopher system we allow all rotations of PH and all rotations of CS. However, for a given symmetry we want to use the "same" rotation for CS as we use for PH. To obtain this we define an atomic colour set INDEX, for which we allow all rotations. PH and CS are then defined as structured coloured sets, obtained from INDEX by means of the union constructor. This means we have the following colour set declarations (where n is a constant of type integer):

```
color INDEX = int with 1..n symgroup rotation
color PH = union ph:INDEX
color CS = union cs:INDEX
```

There are several reasons for introducing permutation symmetries. Firstly, they are rather straightforward and natural. This means they are easy to understand and use. Secondly, it is possible for the user to specify the allowed symmetries at a high abstraction level – as indicated by the examples above. The detailed implementation (e.g., of EquivMark and EquivBE) can then be done automatically by the occurrence graph tool. Thirdly, it is often possible to make a very efficient calculation of permutation symmetries, because sets of colour symmetries can be represented in a compact form, which allows them to be efficiently manipulated. Finally, it is much easier to support the check of consistency properties, which are structural properties (instead of dynamic properties). We shall return to the last three of these reasons in Sect. 3.5, when we discuss computer support of OS-graphs.

Proof rules for permutation symmetries

Permutation symmetries constitute a particular case of the general symmetries defined in Sect. 3.1. Hence it is obvious that we can use the proof rules in Sect. 3.2. However, for the boundedness, liveness, and fairness properties, it is possible to formulate more specific proof rules, because we know more about how markings are mapped into markings and binding elements into binding elements.

When S is a colour set C(p) or a set of bindings B(t), we use Φ to obtain an equivalence relation on S. Two elements $s, s^* \in S$ are equivalent iff there exists a symmetry $\phi \in \Phi$ such that $s = \phi(s^*)$. The equivalence classes of S are denoted by $S_{\approx}$. Finally, we use V_R to denote an arbitrary set of markings which contains (at least) one marking from each equivalence class in V.

Proposition 3.18: For the **boundedness properties** we have the following proof rules, valid for all $p \in PI$:

(i) $\text{BestUpperIntegerBound}(p) = \max_{M \in V_R} |M(p)|$.

(ii) $\text{BestUpperMulti-setBound}(p) = \sum_{C \in C(p)_{\approx}} (\max_{M \in V_R, c \in C} M(p)(c)) * C$.

(iii) $|C(p)_{\approx}| = 1 \Rightarrow$
$\text{BestUpperMulti-setBound}(p) = (\max_{M \in V_R, c \in C(p)} M(p)(c)) * C(p)$.

By replacing max with min, we get three rules for lower bounds.

Explanation: Property (i) is similar to property (ii) in Prop. 3.8. However, it is now sufficient to check a single marking from each equivalence class in V. Property (ii) says we can determine the best upper multi-set bound by calculating, for each equivalence class $C \in C(p)_{\approx}$, the largest coefficient in V_R. Finally, property (iii) investigates the situation in which $C(p)_{\approx}$ has only one equivalence class, i.e., where all colours in C(p) can be mapped into each other.

Proposition 3.18 (ii) shows that the existence of a consistent permutation symmetry specification implies that the best multi-set bounds of the CP-net must be symmetrical, in the sense that all members of an equivalence class in $C(p)_{\approx}$ have the same coefficient.

Proof: Property (i) follows from Prop. 3.8 (ii) and the observation that $M_1 \approx_M M_2$ implies $|M_1(p)| = |M_2(p)|$. Property (ii) follows from Prop. 3.8 (iii) and the observations that $[M_0\rangle = [V] = \Phi(V_R)$ and $(\phi(M))(p,c) = M(p,\phi(c))$ for all $M \in \mathbb{M}$ and all $\phi \in \Phi$. Property (iii) is a particular case of (ii).

□

Next we consider the liveness properties. We use $(t,b) \in c$ to denote that the binding element (t,b) appears in the strongly connected component c, i.e., that (t,b) appears in one of the equivalence classes attached to an arc of c. Analogously, we use $t \in A$ to denote that the transition t appears in the OS-graph, i.e., that t appears in one of the equivalence classes attached to an arc in A.

Proposition 3.19: For the **liveness properties** we have the following proof rules, valid for all $t \in T \cup TI$:

(i) t is live $\Leftrightarrow \forall c \in SCC_T\text{: } t \in BE(c)$.
(ii) t is strictly live $\Leftarrow \forall c \in SCC_T\text{: } [\forall b \in B(t)\text{: } (t,b) \in c \wedge Sym(c)]$.
(iii) $|SCC| = 1 \Rightarrow$ (t is live $\Leftrightarrow t \in A$).
(iv) $|B(t)_{\approx}| = 1 \Rightarrow$ (t is strictly live $\Leftarrow \forall c \in SCC_T\text{: } [t \in BE(c) \wedge Sym(c)]$).
(v) $|B(t)_{\approx}| = |SCC| = 1 \Rightarrow$ (t is strictly live $\Leftrightarrow t \in A$).

Explanation: The first property is similar to property (v) in Prop. 3.10. However, it is now known that [BE(t)] is always identical to BE(t), and hence we can drop this requirement. Property (ii) tells us that a transition is strictly live iff each terminal strongly connected component contains all the binding elements of t and contains an equivalence class with only one element.

Properties (iii)–(v) are particular cases of (i) and (ii). They investigate the situations in which SCC and/or $B(t)_{\approx}$ are known to have only one element. Note that when both properties are satisfied, t is strictly live iff it is live.

Proof: Property (i) follows from Prop. 3.10 (v) by the observation that each symmetry maps a binding element of t into a binding element of t. Property (ii) follows from Prop. 3.10 (vi) and the definition of strict liveness. Property (iii) is a particular case of (i), while (iv) and (v) are particular cases of (ii). □

Finally, let us consider the fairness properties. For a transition or a transition instance $t \in T \cup TI$ we construct a subgraph of the OS-graph by deleting all arcs that contain an equivalence class [b] in which t appears (if t appears in one of the binding elements in [b] it will appear in all of them). We use $SCC_{OS\backslash t}$ to denote the set of all strongly connected components of this subgraph. We also use the notation defined above Prop. 2.9.

Proposition 3.20: For the **fairness properties** we have the following proof rules, valid for all $t \in T \cup TI$:

(i) t is impartial $\Leftrightarrow$ $\forall dc \in DCS$: $[BE(dc) \cap BE(t) \neq \emptyset]$.
(ii) t is fair $\Leftrightarrow$ $\forall dc \in DCS$: $[BE(dc) \cap BE(t) \neq \emptyset \vee \forall C \in dc$: $BE(C) \cap BE(t) = \emptyset]$.
(iii) t is just $\Leftrightarrow$ $\forall dc \in DCS$: $[BE(dc) \cap BE(t) \neq \emptyset \vee \exists C \in dc$: $BE(C) \cap BE(t) = \emptyset]$.

(iv) t is impartial $\Leftrightarrow$ $\forall c \in SCC_{OS\backslash t}$: [c is trivial].
(v) t is fair $\Leftrightarrow$ $\forall c \in SCC_{OS\backslash t}$: [c is trivial $\vee$ $\forall C \in c$: $BE(C) \cap BE(t) = \emptyset$].
(vi) t is just $\Leftrightarrow$ $\forall c \in SCC_{OS\backslash t}$: [c is trivial $\vee$
$\forall dc \in DCS(c)$ $\exists C \in dc$: $BE(C) \cap BE(t) = \emptyset$].

Explanation: All six properties are particular cases of the properties in Prop. 2.10. We simply set $X = BE(t)$.

Proof: Immediate consequence of Prop. 2.10 and the observation that $[BE(t)] = BE(t)$. □

3.4 Labelled OS-graphs

In this section we define labelled OS-graphs. The ideas behind this kind of graph are analogous to the ideas behind labelled OE-graphs. By adding node and arc labels, we make it possible to record information about individual markings and individual binding elements.

In a labelled OE-graph each arc label is a set which consists of a number of pairs $(b,M_2) \in BE \times \mathbb{M}$ such that $M_1 [b\rangle M_2$ where M_1 is the label of the source node and M_2 is known to be equivalent to the label M_D of the destination node. We use that same idea for labelled OS-graphs. However, now we have a symmetry specification and hence we know that $M_2 \approx_M M_D$ implies the existence of a symmetry $\phi \in \Phi$ such that $M_2 = \phi(M_D)$. Hence we can record ϕ instead of M_2. This means each arc label consists of a number of pairs $(b,\phi) \in BE \times \Phi$. It is often possible to represent a symmetry much more succinctly than a marking. Hence we save space in the representation of the OS-graph. The above idea was already used in Sect. 2.6, where the arc labels in Fig. 2.7 contain pairs of the form (b,ϕ). This was possible because the equivalence specification for the data base system can be derived from a symmetry specification.

A symmetry ϕ that maps a marking M into itself is said to be a **self-symmetry** of M. We use $\Phi_M \subseteq \Phi$ to denote the set of all self-symmetries of M:

$$\Phi_M = \{\phi \in \Phi \mid \phi(M) = M\}.$$

It is easy to prove that Φ_M is an algebraic subgroup of Φ. A marking for which $\Phi_M = \Phi$ is said to be **symmetrical**. In Sect. 3.5 we discuss how self-symmetries can be calculated, and we also show that self-symmetries allow us to obtain a more efficient version of the algorithm in Prop. 2.5.

In a labelled OS-graph we introduce a new labelling function Φ which maps each node $v \in V$ into the set of self-symmetries of M_v (i.e., the self-symmetries of the marking attached to v by the labelling function M). This allows us to obtain a significant reduction of the arc labels. The reason is that the consistency property in Def. 3.3 (ii) tells us that $M_1 [b\rangle M_2$ implies:

(*) $\quad M_1 = \phi(M_1) [\phi(b)\rangle \phi(M_2)$

for all self-symmetries $\phi \in \Phi_{M_1}$. All the occurrences in (*) involve markings and binding elements that belong to the same three equivalence classes: $[M_1]$, $[b]$, and $[M_2]$. Hence they are all represented by the same arc. For this arc it is sufficient that the arc label contains the pair (b,ϕ^*) corresponding to the occurrence $M_1 [b\rangle M_2$. None of the other occurrences of the form in (*) need to be explicitly represented in the arc label, because the pair (b',ϕ'), corresponding to $\phi(M_1) [\phi(b)\rangle \phi(M_2)$, can be computed from (b,ϕ^*) as follows:

$$(b',\phi') = (\phi(b), \phi \circ \phi^*).$$

Now we are ready to define labelled OS-graphs. Analogously to labelled OE-graphs, we denote the labelling functions by M_V, Φ_V and L_a. Some explanation follows the definition:

Definition 3.21: Let a CP-net and a consistent symmetry specification Φ be given. A **labelled OS-graph** is a tuple OSLG = (OSG, M, Φ, L) where OSG = (V, A, N) is the OS-graph, while $M \in [V \to \mathbb{M}]$, $\Phi \in [V \to \Phi_S]$, and $L \in [A \to (BE \times \Phi)_S]$ are **labelling** functions satisfying the following properties:

(i) $\forall v \in V$: $M_v \in v$.
(ii) $\forall v \in V$: $\Phi_v = \Phi_{M_v}$.
(iii) $\forall a \in A$: $[Next(M_{s(a)}) \cap (BE \times d(a)) = \cup_{(b,\phi) \in L_a} \Phi_{s(a)}(b,\phi(M_{d(a)})) \wedge$
$\forall (b',\phi'),(b'',\phi'') \in L_a$: $b' \in \Phi_{s(a)}(b'') \Rightarrow (b',\phi') = (b'',\phi'')]$.

Properties (i) and (ii) of Def. 3.21 are straightforward. Property (iii) is a bit more complex. It consists of two parts. The first part guarantees that the arc label L_a is large enough, in the sense that each pair:

$$(b,M) \in Next(M_{s(a)}) \cap (BE \times d(a))$$

can be obtained by using a self-symmetry $\phi \in \Phi_{s(a)}$ of the source node to a pair $(b,\phi(M_{d(a)}))$ where $(b,\phi) \in L_a$. The second property guarantees that L_a is minimal, in the sense that it does not contain two pairs (b',ϕ') and (b'',ϕ'') whose binding elements can be obtained from each other by means of a self-symmetry of the source node.

It is obvious that a given CP-net and symmetry specification may have many different labelled OS-graphs (since we can choose the marking labels and the arc labels in many different ways).

Each labelled OS-graph contains more detailed information than the corresponding OS-graph. However, it is not more difficult or more time-consuming to calculate the labelled OS-graph. The symmetries in the arc labels have to be calculated under all circumstances, to determine which equivalence class the new marking belongs to. In the labelled OS-graph we record the symmetry. In the OS-graph we just throw it away. Analogously, the self-symmetries are calculated in all circumstances, because they can be used to improve the efficiency of the algorithm in Prop. 2.5. For more details see Sect. 3.5.

Labelled OS-graphs for philosopher systems with five and nine philosophers are shown in Figs. 3.4 and 3.5. In the arc labels each line represents a pair (for which we omit the surrounding parenthesis). We use *Ti* and *Pi* as shorthand for (TakeChopsticks, <p=ph(i)>) and (PutDownChopsticks, <p=ph(i)>), respectively. We omit symmetries that are the identity function. Moreover, we draw all arcs as double arcs and hence we lump the two corresponding arc labels into a single label. As an example, the double arc between nodes #2 and #3 of Fig. 3.4 represents two arcs, one in each direction. The first two lines of the arc label

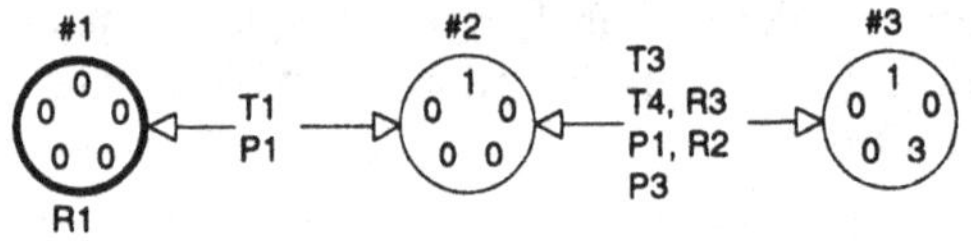

#1 #2 #3
T1
P1
T3
T4, R3
P1, R2
P3
R1

Fig. 3.4. Labelled OS-graph for five philosophers

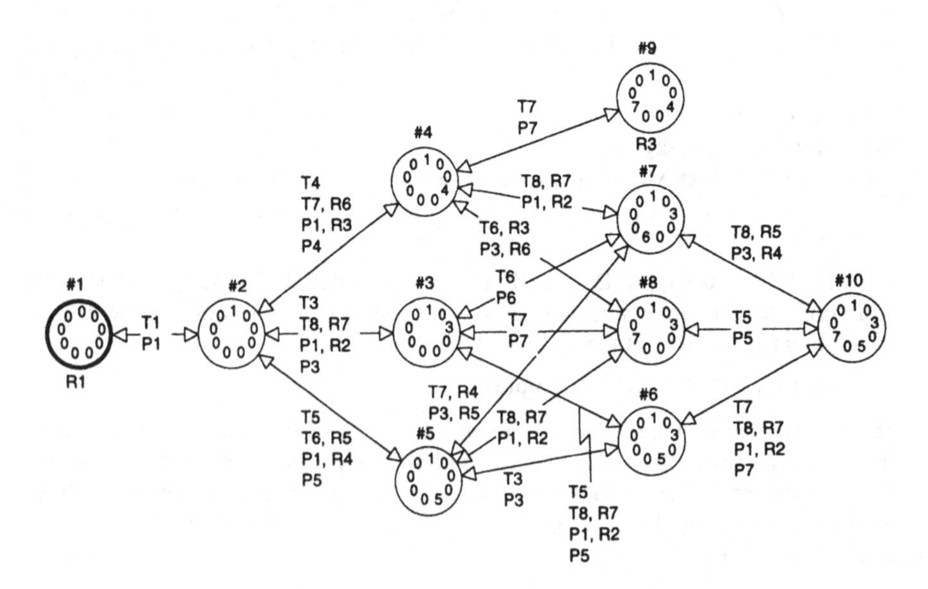

Fig. 3.5. Labelled OS-graph for nine philosophers

belong to the arc from #2 to #3. The first line tells us that T3 is enabled in node #2 and will lead to the marking of node #3. The second line tells us that T4 is enabled in node #2 and will lead to a marking that can be obtained from node #3 by the symmetry R_3 (which denotes a rotation of length 3; see Sect. 3.1). The last two lines belong to the arc from #3 to #2. They tell us that P1 and P3 are enabled in node #3. P1 will lead to a marking that can be obtained from node #2 by the symmetry R_2, while P3 will lead to the marking of node #2.

The self-symmetry labels are shown below each node (a missing label indicates that only the identity function is a self-symmetry). Here, we exploit the fact that Φ_R is a finite **cyclic group**. This means there exists an element $\phi \in \Phi_R$, called the generator, and a positive integer $n \in \mathbb{N}_+$ called the order, such that Φ_R can be written in the following form:

$$\Phi_R = \{\phi^k \mid k \in 1..n\}$$

where ϕ^k denotes composition of ϕ with itself k times and $\phi^i = \phi^j \Rightarrow i = j$ for all $i,j \in 1..n$. A subgroup of a cyclic group is again a cyclic group, and hence we can denote each set of self-symmetries by its generator, i.e., by a single symmetry.

For node #1 in Fig. 3.4, the generator is R_1. This means all rotations are self-symmetries. The two other nodes in Fig. 3.4 have R_5 as generator. This means only the identity function is a self-symmetry (and hence the self-symmetry labels are omitted). For node #9 of Fig. 3.5, the generator is R_3. This means the set of self-symmetries is $\{R_3, R_6, R_9\}$.

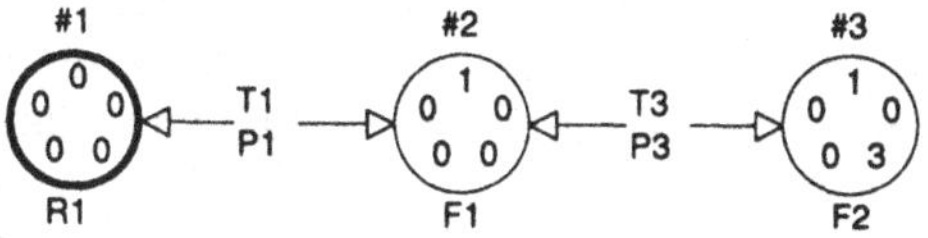

Fig. 3.6. Labelled OS-graph for five philosophers (with flip operations)

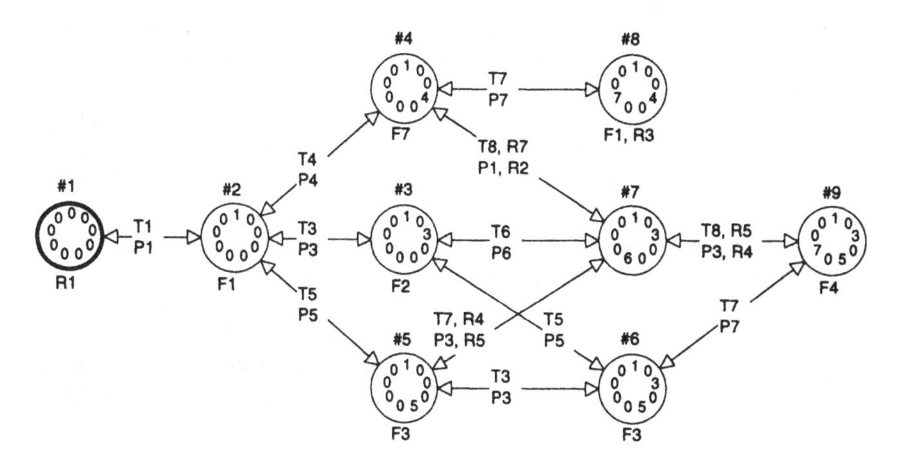

Fig. 3.7. Labelled OS-graph for nine philosophers (with flip operations)

For the philosopher system we can obtain an alternative symmetry specification by allowing rotations to be supplemented by flip operations. For $k \in 1..n$ we define F_k to be the permutation that maps each k into itself and interchanges k+1 with k–1, k+2 with k–2, etc. (the plus and minus signs denote cyclic operations). We now define:

$$\Phi_{RF} = \{R_k \mid k \in 1..n\} \cup \{F_k \mid k \in 1..n\}.$$

It is then easy to verify that Φ_{RF} determines a consistent symmetry specification. Each F_k is its own inverse.

By means of Φ_{RF} we obtain the labelled OS-graphs in Figs. 3.6 and 3.7. The new graphs have smaller arc labels than the old ones and they also have fewer nodes and arcs (when there are nine or more philosophers). However, it is not so easy to judge whether Φ_{RF} also gives an improved time complexity of the algorithm in Prop. 2.5. On the one hand, we do not have to calculate as many occurring binding elements as before. On the other hand, it becomes significantly more difficult to calculate the set of self-symmetries (since we no longer have a cyclic group). This means we sometimes need more than one generator (see node #8 in Fig. 3.7). It also becomes more difficult to test whether two markings are equivalent or not, because we also have to consider all the members of $\{F_k \mid k \in 1..n\}$.

From the discussion in this chapter, it follows that labelled OS-graphs are more powerful than ordinary OS-graphs, although they are not more difficult to construct. Hence it is obvious to ask why we have introduced unlabelled OS-graphs at all. There are several reasons for this. Firstly, there are a lot of situations in which we do not need the labels, because the proof rules in Sect. 3.2 are sufficient. Secondly, the definition of the OS-graph labels introduce self-symmetries, i.e., a new non-trivial concept which the user has to learn. Self-symmetries are used to optimise the construction of OS-graphs, but for unlabelled OS-graphs the user does not need to know anything about them in order to understand and interpret the graphs. Thirdly, unlabelled OS-graphs are directly based on unlabelled OE-graphs. The only new concept the user has to learn is that of a symmetry specification. Finally, a labelled OS-graph may use considerably more space than an ordinary OS-graph, because the arc labels and the set of self-symmetries may be rather large.

The following theorem shows that it is possible to construct the O-graph of a CP-net from any of the labelled OS-graphs. This means each OS-graph contains the same amount of information as the O-graph. Hence we conclude that labelled OS-graphs can be used to investigate any property that can be investigated by means of O-graphs. We may consider a labelled OS-graph to be a condensed version of the O-graph – rather as a CP-net is a condensed version of the equivalent PT-net.

In practice we never construct an O-graph from a labelled OS-graph. This means Theorem 3.22 only has theoretical importance. Hence it may be skipped by readers who are primarily interested in the practical application of CP-nets.

Theorem 3.22: Let $OSGL^* = ((V^*, A^*, N^*), M, \Phi, L)$ be a labelled OS-graph for a CP-net, CPN. The O-graph of CPN is the directed graph $OG' = (V', A', N')$ where:

(i) $V' = [V^*]$.

(ii) $A' = \{\phi(M_{s(a^*)}, b^*, \phi^*(M_{d(a^*)})) \mid a^* \in A^* \wedge (b^*, \phi^*) \in L_{a^*} \wedge \phi \in \Phi\}$.

(iii) $\forall a=(M_1,b,M_2) \in A'$: $N'(a) = (M_1,M_2)$.

Proof: We have to prove that OG' is identical to the directed graph OG = (V, A, N) defined in Def. 1.3. We first prove that V' = V. This follows from Prop. 1.12 (i) and Prop. 3.7 (i).

Next we prove that A' = A. This is done by proving two inclusions. First we prove that $A' \subseteq A$. Assume that we have an arc:

$$a = \phi(M_{s(a^*)}, b^*, \phi^*(M_{d(a^*)})) \in A'$$

where $a^* \in A^*$, $(b^*, \phi^*) \in L_{a^*}$ and $\phi \in \Phi$. From $a^* \in A^*$, $(b^*, \phi^*) \in L_{a^*}$ and the first part of Def. 3.21 (iii) we know that:

$$(b^*, \phi^*(M_{d(a^*)})) \in Next(M_{s(a^*)}).$$

Hence we have:

$$M_{s(a^*)} [b^*\rangle\, \phi^*(M_{d(a^*)}),$$

and by Def. 3.3 (ii) we get:

$$\phi(M_{s(a^*)})\ [\phi(b^*)\rangle\ (\phi \circ \phi^*)(M_{d(a^*)}),$$

which by Def. 1.3 (ii) implies that $a \in A$.

Secondly, we prove that $A \subseteq A'$. Assume that we have an arc:

$$a = (M_1, b, M_2) \in A.$$

From $a \in A$ and Def. 1.3 (i) we know that $M_1, M_2 \in [M_0\rangle$ and from Def. 1.3 (ii) we know that:

$$M_1 [b\rangle M_2.$$

From (i) and Def. 3.21 (i) we know there exist a node $v^* \in V^*$ and a symmetry ϕ_1 such that $M_{v^*} = \phi_1(M_1)$. From Def. 3.3 (ii) we then conclude that:

$$M_{v^*} = \phi_1(M_1) [\phi_1(b)\rangle\, \phi_1(M_2).$$

From Def. 2.3 (ii) we then know there exists an arc:

$$a^* = ([\phi_1(M_1)], [\phi_1(b)], [\phi_1(M_2)]) \in A^*,$$

and from the first part of Def. 3.21 (iii) we know there exist a pair $(b^*, \phi^*) \in L_{a^*}$ and a self-symmetry $\phi_2 \in \Phi_{v^*}$ such that:

(*) $\quad (\phi_1(b), \phi_1(M_2)) = \phi_2(b^*, \phi^*(M_{d(a^*)})).$

From $M_{v^*} = \phi_1(M_1)$ and $\phi_2 \in \Phi_{v^*}$ we have:

(**) $\quad M_1 = (\phi_1^{-1} \circ \phi_2)(M_{v^*}) = (\phi_1^{-1} \circ \phi_2)(M_{s(a^*)}).$

From (*) and (**) we conclude that:

$$(M_1,b,M_2) = (\phi_1^{-1} \circ \phi_2)(M_{s(a*)}, b^*, \phi^*(M_{d(a*)})),$$

and from (ii) we then conclude that $a \in A'$.

Finally, we have to prove that N' = N. This is a direct consequence of (iii) and Def. 1.3 (iii). □

3.5 Computer Tools for OS-graphs

This section describes how the construction and analysis of OS-graphs are supported by computer tools. In Sects. 1.7 and 2.7 we discussed tool support for O-graphs and OE-graphs. It should be obvious that most of that discussion is also valid for tools that deal with OS-graphs. In the present section we shall not repeat the arguments of Sects. 1.7 and 2.7. Instead we focus on the additional requirements needed to support OS-graphs.

The only difference between an OE-graph and an OS-graph is the fact that the equivalence specification in the OS-graph is defined via a symmetry specification. This means we have a stronger consistency requirement (Def. 3.3 instead of Def. 2.2).

Analysis of OS-graphs

The analysis of OS-graphs can be done in exactly the same way as the analysis of O-graphs and OE-graphs, i.e., by means of SearchNodes, SearchArcs, and SearchComponents. This is the case because the right-hand sides of the proof rules in Sect. 3.2 are similar to the right-hand sides of the proof rules in Sects. 1.4 and 2.3 (except for small notational differences). The main difference is that, for OS-graphs, we want to find those nodes where the equivalence class has only one element. This can be done during the construction of the OS-graph when the self-symmetries are calculated.

Construction of OS-graphs

It is possible to construct OS-graphs in a more efficient way than we can construct OE-graphs. This can be seen by taking a closer look at the for-loop in Prop. 2.5. Assume that, for a marking M_1, we have processed a binding element b such that:

$$M_1\,[b\rangle\, M_2.$$

Then we do not have to process any of those binding elements which are of the form $\phi(b)$ where $\phi \in \Phi_{M_1}$. The reason is that the consistency property in Def. 3.3 (ii) and the definition of self-symmetries tell us that:

$$M_1 = \phi(M_1)\,[\phi(b)\rangle\,\phi(M_2).$$

This means the processing of the binding element $\phi(b)$ will add no new nodes or arcs to the OS-graph, because $\phi(M_2) \approx_M M_2$ and $\phi(b) \approx_{BE} b$. Hence we do not need to calculate the successor that corresponds to $\phi(b)$, and (more importantly) we save the test whether the successor marking is equivalent to any of the existing markings in the OS-graph. For a labelled OS-graph the situation is similar, because the second part of Def. 3.21 (iii) guarantees that the processing of $\phi(b)$ cannot add to the arc label.

Self-symmetries may also be used to optimise the test whether two markings M_1 and M_2 are equivalent or not. A **left coset** of the subgroup Φ_{M_1} is a set of the form:

$$\phi^* \circ \Phi_{M_1} = \{\phi^* \circ \phi \mid \phi \in \Phi_{M_1}\}$$

where $\phi^* \in \Phi$. From group theory we know that two left cosets (of the same subgroup) are either identical or mutually disjoint. This means the set of all left cosets of Φ_{M_1} is a partition of Φ. From the definition of Φ_{M_1} it follows that $\phi'(M_1) = \phi''(M_1)$ for any two symmetries that belong to the same coset of Φ_{M_1}. Hence we can determine the set of symmetries that map M_1 into M_2:

$$\Phi_{M_1,M_2} = \{\phi \in \Phi \mid \phi(M_1) = M_2\}$$

by investigating only one symmetry from each coset. This means we have to investigate $|\Phi|/|\Phi_{M_1}|$ symmetries, instead of $|\Phi|$. When M_1 is symmetrical (i.e., $\Phi_{M_1} = \Phi$) we only have to investigate a single symmetry.

The expense for the above optimisations is that we have to calculate self-symmetries. From the coset partition considered above, we conclude that the size of Φ_{M_1} divides the size of Φ. This can be used to optimise the calculation of Φ_{M_1}.

As an example, consider the case where the size of Φ is a prime. Then there are only two subgroups – the one which is identical to Φ and the one which only contains the neutral element. Hence we can find Φ_{M_1} by investigating a single symmetry (it does not matter which one we choose, as long as we avoid the neutral element).

As another example, consider the case where Φ is known to be a finite cyclic group with generator ϕ and order n (see Sect. 3.4). In this case, it is known that all subgroups of Φ are cyclic, and that each of them has a generator ϕ^k such that k divides n. When n = 100 there are nine different subgroups, with the following generators: ϕ^1, ϕ^2, ϕ^4, ϕ^5, ϕ^{10}, ϕ^{20}, ϕ^{25}, ϕ^{50}, ϕ^{100}. To determine the set of self-symmetries of a given marking, we test each generator (in the above order). The first generator which is a self-symmetry generates the subgroup of all self-symmetries. Hence we have to test at most eight symmetries (there is no need to test ϕ^{100} which is the identity function). For n = 1000 we have to test at most 15 symmetries.

Even when Φ is not cyclic, we may use an idea which is similar to the one described above. Whenever we have found an element $\phi \in \Phi_M$, we know that the cyclic subgroup $G(\phi)$ generated by ϕ is a subset of Φ_M. Hence we do not have to check the remaining symmetries in $G(\phi)$. Neither do we have to test any symmetry which belongs to the left coset $\phi^* \circ G(\phi)$ or the right coset $G(\phi) \circ \phi^*$ of a symmetry ϕ^* which we already know belongs to Φ_M.

Above we have only sketched some of the ways in which self-symmetries and group theory can be used to improve the efficiency of the algorithm in Prop. 2.5.

As for OE-graphs we may leave the implementation of the equivalence tests to the user. This means the user has to write three SML functions, **SelfSym**, **EquivMark** and **EquivBE**. The first function takes a marking M and returns the set of self-symmetries Φ_M. The second function takes two markings M_1 and M_2 and tells us whether they are equivalent or not (using the self-symmetries to optimise the calculation). Analogously, the third function takes two binding elements b_1 and b_2 and tells us whether they are equivalent or not.

We may use self-symmetries to improve the keys used in marking records (see Sect. 1.7). It is rather straightforward to prove that:

$$\Phi_{M_1,M_2} \neq \emptyset \;\Rightarrow\; |\Phi_{M_1}| = |\Phi_{M_2}| = |\Phi_{M_1,M_2}|.$$

This means two equivalent markings always have the same number of self-symmetries. Hence it may be a good idea to use the size of the set of self-symmetries in the function that constructs the keys of the marking records.

Finally, we can support a check of the consistency properties in Def. 3.3, in a similar way as we support a check of the soundness properties of OE-graphs. For a concrete binding element b and two markings M_1 and M_2 (given by the user) the tool can test whether Def. 3.3 (ii) is satisfied for all symmetries $\phi \in \Phi$. Analogously, the tool can check whether the initial marking M_0 is symmetrical or not.

Permutation symmetries

To make a permutation symmetry specification the user simply defines a symmetry group for each atomic colour set. For the three most common kinds of symmetry groups, this can be done very easily. The user simply attaches one of the keywords permutation, rotation, and fixed to the declaration of the atomic colour set – as indicated by the examples in Sect. 3.3. Based on this information the occurrence graph tool may automatically generate the code necessary to calculate symmetries and self-symmetries. A prototype of this exists.

During the calculation the tool represents sets of symmetries in a very compact way, which allows efficient manipulations of them. To illustrate the basic ideas behind the representation, let us assume that we have a CP-net which contains a place instance p* with an atomic colour set S = {a,b,c,d,e,f,g,h,i,j} for which we allow all permutations. It is easy to see that two multi-sets in S_{MS} can be mapped into each other by a symmetry iff they have the same multi-set of positive coefficients. For the multi-sets:

M_1(p*) = 2\`a+3\`b+3\`d+1\`e+2\`f+2\`h

M_2(p*) = 3\`a+2\`c+1\`d+2\`g+2\`h+3\`j

this is the case. Both have 1\`1+3\`2+2\`3 as the coefficient multi-set. The set of all symmetries that map M_1(p*) into M_2(p*) is represented as follows:

$$(*)\quad \begin{bmatrix} e & d \\ afh & cgh \\ bd & aj \end{bmatrix}$$

Each line in (*) describes a property that a symmetry ϕ must fulfil in order to map $M_1(p^*)$ into $M_2(p^*)$. The first line tells us that ϕ must map e into d (because these are the only colours that have 1 as coefficient). Analogously, the second line tells us that ϕ must map the elements of {a,f,h} into the elements of {c,g,h} (because these are the only colours that have 2 as coefficient). Finally, the third line tells us that the elements of {b,d} must be mapped into the elements of {a,j} (because these are the only colours that have 3 as coefficient). The mapping of {a,f,h} into {c,g,h} can be done in 6 different ways (since ϕ is a permutation). Analogously, the mapping of {b,d} into {a,j} can be done in 2 different ways, while the four colours {c,g,i,j} that are absent in $M_1(p^*)$ can be mapped into the four colours {b,e,f,i} that are absent in $M_2(p^*)$ in 24 different ways. Hence we conclude that there are 288 different symmetries that map $M_1(p^*)$ into $M_2(p^*)$. All these symmetries are represented by (*), which is called a **restriction set**, because each line describes a **restriction** that the symmetries must fulfil.

Now let us assume that the CP-net has another place instance p** that also has S as colour set and let us assume that p** has the following multi-sets:

$$M_1(p^{**}) = 1`a + 2`f$$

$$M_2(p^{**}) = 2`c + 1`g.$$

Then we get the following restriction set:

$$(**)\quad \begin{bmatrix} a & g \\ f & c \end{bmatrix},$$

which represents $8! = 40{,}320$ different symmetries. Now let us try to find the symmetries that map $M_1(p^*)$ into $M_2(p^*)$ and $M_1(p^{**})$ into $M_2(p^{**})$, i.e., the symmetries that satisfy both (*) and (**). To do this we calculate the intersection of the symmetry sets represented by (*) and (**). This can be done by a very simple and efficient algorithm, which compares the individual restrictions of (*) and (**). We shall not give the algorithm, but only indicate the result:

$$\begin{bmatrix} e & d \\ afh & cgh \\ bd & aj \end{bmatrix} \cap \begin{bmatrix} a & g \\ f & c \end{bmatrix} = \begin{bmatrix} e & d \\ a & g \\ f & c \\ h & h \\ bd & aj \end{bmatrix}$$

As long as we only consider atomic colour sets (and allow all permutations), it can be proved that Φ_{M_1,M_2} either is empty or can be written as a restriction set (calculated by means of a number of intersection operations).

Unfortunately, the situation is a bit more complicated for some of the structured colour sets: products, records, and lists. To illustrate the problems let us consider a place instance p that has S×S as colour set and let us assume that p has the following multi-sets:

$$M_1(p) = 1`(a,a) + 1`(d,d) + 2`(e,b)$$

$$M_2(p) = 1`(a,d) + 2`(c,h) + 1`(d,a).$$

From Def. 3.12 (i) it is easy to see that a symmetry $\phi \in \Phi_{M_1,M_2}$ must map $PR_1(M_1(p))$ into $PR_1(M_2(p))$ and $PR_2(M_1(p))$ into $PR_2(M_2(p))$, where PR_1 and PR_2 are the two projection functions from S×S into S. This means ϕ must fulfil the following restriction set:

$$\begin{bmatrix} ad & ad \\ e & c \end{bmatrix} \cap \begin{bmatrix} ad & ad \\ b & h \end{bmatrix} = \begin{bmatrix} ad & ad \\ e & c \\ b & h \end{bmatrix}$$

Unfortunately, this is not a sufficient condition, because it does not take into consideration how colours in the first component of S×S are paired with colours in the second. Actually, it is impossible to find a symmetry mapping $M_1(p)$ into $M_2(p)$. This can be seen by observing that it is impossible to find a symmetry mapping the token $(a,a) \in M_1(p)$ into any of the tokens contained in $M_2(p)$.

When we deal with symmetries between two markings M_1 and M_2, which have structured colour sets, we can use restriction sets and projections (plus a number of similar techniques) to obtain an approximation of Φ_{M_1,M_2}, i.e., a set of symmetries $\Phi^*_{M_1,M_2}$ such that $\Phi_{M_1,M_2} \subseteq \Phi^*_{M_1,M_2}$. It is then necessary to test the individual symmetries of $\Phi^*_{M_1,M_2}$ to see whether they really belong to Φ_{M_1,M_2}. Fortunately, there are many situations in which this additional test is unnecessary, e.g., when a structured multi-set only contains one token.

The ideas behind restriction sets can also be used to find self-symmetries. Then it will always be the case that the left-hand side of each restriction is identical to the right-hand side – and hence we omit the latter. As an example, we can represent all the self-symmetries of the multi-set:

$$M_1(p^*) = 2`a + 3`b + 3`d + 1`e + 2`f + 2`h$$

by the following **component set**:

$$\begin{bmatrix} e \\ afh \\ bd \end{bmatrix}$$

Each component set represents a subgroup of the set of all permutations. Hence we can use component sets to specify non-standard symmetry groups. However, not all subgroups can be represented by a component set. This is not possible, e.g., for the set of all rotations.

As mentioned in Sect. 3.4, permutation symmetries have a set of consistency properties that can be checked by structural tests, i.e., without considering the set of all reachable markings. The check of Def. 3.16 (iii) can be done in a way very similar to the check of a proposed place invariant. In both cases, we have to check whether some expression (built from one or more arc expressions) evaluates to the empty multi-set for all possible bindings of the transition. In Sect. 4.4 we discuss how this can be done. Also Def. 3.16 (i) and (ii) can be handled by this method.

Bibliographical Remarks

Many of the concepts introduced in this section are well known from standard algebra. For example, the subgroup of self-symmetries is the isotropy subgroup of the group operation which maps any pair $(\phi,M) \in \Phi \times \mathbb{M}$ into $\phi(M)$.

The use of permutation symmetries to construct OS-graphs (and labelled OS-graphs) is in many respects closely related to the ideas behind symbolic reachability graphs, described in [6] and [8]. Some of the proof rules in Sect. 3.2 are almost identical to some of the proof rules in [8]. Moreover, both methods use self-symmetries to reduce the size of the arc labels. The latter idea has been independently developed by the two research groups involved. The main difference between the two approaches is the fact that symbolic reachability graphs have a symbolic, unique representation of each equivalence class. This simplifies the calculation of symmetries. However, it only works for a restricted set of CP-nets, called well-formed CP-nets. The use of permutation symmetries to reduce occurrence graphs was originally proposed in [21]. Related approaches are also described in [11], [16], and [30]. The first two of these papers deal with arbitrary transition systems, while the third paper deals with Predicate/Transition Nets. A detailed study of the representation and calculation of permutation symmetries can be found in [33].

Another technique for reducing the size of occurrence graphs is presented in [42] and [43]. It builds upon the observation that a CP-net often has a number of occurrence sequences where the steps are identical, except for the order in which they occur. Let a marking M have n concurrently enabled binding elements (which are different). Then we can sort these elements in n! different ways. Each of these determines a possible occurrence sequence starting in M. For $n = 4$ we have 24 different possibilities. For $n = 10$ we have 3,628,800 possibilities. However, the total effects of all these occurrence sequences are the same, in the sense that they all lead to the same marking. Thus it is natural to ask whether it is really necessary to develop all of them, and fortunately it turns out that this is not the case. Instead, for each reachable marking, we calculate a so-called stubborn set, and we use this set to tell us which of the enabled binding elements we need to investigate, i.e., to let occur. The remaining binding elements are, by the definition of stubborn sets, guaranteed to remain enabled, and thus they can occur in the next marking (or in a later one). The use of stubborn sets often gives a very significant reduction of the number of nodes and the number of arcs – in particular when the modelled system contains a large number of relatively independent processes. Unfortunately, with the stubborn set method it is sometimes necessary to construct several different occurrence graphs (because the definition of stubborn sets depends upon those properties which we want to investigate). The use of stubborn sets can be combined with the use of symmetrical markings. The two techniques are orthogonal, in the sense that symmetrical markings are useful when we have a number of symmetrical processes, while stubborn sets are useful when we have a number of concurrent processes. When the processes are both symmetrical and concurrent we can use both techniques simultaneously. For details see [40].

A third reduction possibility is described in [18] and [28]. It involves looking for occurrence sequences that lead from a reachable marking M_1 to a covering marking M_2 (i.e., a marking which is strictly larger than M_1). The total effect of such a sequence is to add tokens, and thus it is easy to see that the steps can be repeated as many times as we want. This means some of the token elements can get an arbitrarily high coefficient, because each repetition of the sequence increases the coefficient. Such coefficients are replaced by ∞ and we then have an occurrence graph where some nodes represent many different markings (which are all identical except for those token elements which have a ∞ coefficient). For a CP-net with finite colour sets it can be proved that reduction by means of covering markings always yields a finite occurrence graph. However, the technique has two main drawbacks. Firstly it only gives a reduction for unbounded systems (and most practical systems are bounded). And secondly, so much information is lost by the reduction that several important properties (e.g., liveness and reachability) are no longer fully decidable. In [21] and [34] it is shown how to use covering markings together with permutation symmetries. The theoretical justification of the combined technique becomes rather complex.

Exercises

Exercise 3.1.
Consider the data base system from Sect. 1.3 of Vol. 1 and the OS-graphs in Figs. 2.1 and 2.2.

(a) Prove that the symmetry specification allowing all permutations of DBM is consistent, i.e., fulfils the properties in Def. 3.3.

(b) Investigate to what extent the proof rules of Prop. 3.8 and Prop. 3.18 are superior to those of Prop. 2.7 with respect to verification of the upper and lower bounds postulated at the end of Sect. 4.2 of Vol. 1.

(c) Investigate to what extent the proof rules of Prop. 3.9 are superior to those of Prop. 2.8 with respect to verification of the home properties postulated at the end of Sect. 4.3 of Vol. 1.

(d) Investigate to what extent the proof rules of Prop. 3.10 and Prop. 3.19 are superior to those of Prop. 2.9 with respect to verification of the liveness properties postulated at the end of Sect. 4.4 of Vol. 1.

Exercise 3.2.
Consider the philosopher system from Sect. 1.6 and the OS-graphs in Figs. 3.1 and 3.2.

(a) Prove that the symmetry specification allowing all rotations is consistent, i.e., fulfils the properties in Def. 3.3.

(b) Prove that the symmetry specification allowing all permutations is inconsistent, i.e., violates the properties in Def. 3.3.

(c) Use the OS-graphs and the proof rules in Prop. 3.8 and Prop. 3.18 to investigate the boundedness properties of the philosopher system.

(d) Use the OS-graphs and the proof rules in Prop. 3.9 to investigate the home properties of the philosopher system.

(e) Use the OS-graphs and the proof rules in Prop. 3.10 and Prop. 3.19 to investigate the liveness properties of the philosopher system.

(f) Use the OS-graphs and the proof rules in Prop. 2.10 and Prop. 3.20 to investigate the fairness properties of the philosopher system.

Exercise 3.3.
Consider the telephone system from Sect. 3.2 of Vol. 1.

(a) Prove that the symmetry specification defined in Sect. 3.3 is consistent, i.e., fulfils the properties in Def. 3.3

(b) Use the symmetry specification to obtain an OS-graph for the telephone system with $|U| = 3$. What is the size of the OS-graph?

(c) Investigate whether you can use the OS-graph and the proof rules in Prop. 3.8 and Prop. 3.18 to verify the upper and lower bounds postulated at the end of Sect. 4.2 of Vol. 1.

(d) Investigate whether you can use the OS-graph and the proof rules in Prop. 3.9 to verify the home properties postulated at the end of Sect. 4.3 of Vol. 1.

(e) Investigate whether you can use the OS-graph and the proof rules in Prop. 3.10 and Prop. 3.19 to verify the liveness properties postulated at the end of Sect. 4.4 of Vol. 1. Investigate whether any of the transitions are strictly live.

(f) Investigate whether you can use the OS-graph and the proof rules in Prop. 2.10 and Prop. 3.20 to verify the fairness properties postulated at the end of Sect. 4.5 of Vol. 1.

(g) Repeat (b)–(f) with $|U| = 4$ and with $|U| = 5$ (if possible).

Exercise 3.4.
Consider the process control system from Exercise 4.6 of Vol. 1. This exercise is only worth attempting if you have access to an occurrence graph tool. Even though the net is rather small, it will take too long to produce the OS-graphs if this has to be done manually.

(a) Prove that the symmetry specification defined in Sect. 3.3 is consistent, i.e., fulfils the properties in Def. 3.3

(b) Use the symmetry specification to obtain an OS-graph for the process control system with $|PROC| = |RES| = 2$. What is the size of the OS-graph?

(c) Use the OS-graph and the proof rules in Prop. 3.8 and Prop. 3.18 to investigate the boundedness properties of the process control system.

(d) Use the OS-graph and the proof rules in Prop. 3.9 to investigate the home properties of the process control system.

(e) Use the OS-graph and the proof rules in Prop. 3.10 and Prop. 3.19 to investigate the liveness properties of the process control system.

(f) Use the OS-graph and the proof rules in Prop. 2.10 and Prop. 3.20 to investigate the fairness properties of the process control system.

(g) Repeat (b)–(f) with $|PROC| = |RES| = 3$ and with $|PROC| = |RES| = 4$ (if possible).

Exercise 3.5.
Consider the modified ring network from Exercise 1.6. This exercise is only worth attempting if you have access to an occurrence graph tool. Even though the net is rather small, it will take too long to produce the OS-graphs if this has to be done manually.

(a) Prove that the symmetry specification defined in Sect. 3.3 is consistent, i.e., fulfils the properties in Def. 3.3.

(b) Use the symmetry specification to obtain an OS-graph for the ring network with NoOfBuffers = 2. What is the size of the OS-graph?

(c) Use the OS-graph and the proof rules in Prop. 3.8 and Prop. 3.18 to investigate the boundedness properties of the modified ring network.

(d) Use the OS-graph and the proof rules in Prop. 3.9 to investigate the home properties of the modified ring network.

(e) Use the OS-graph and the proof rules in Prop. 3.10 and Prop. 3.19 to investigate the liveness properties of the modified ring network.

(f) Use the OS-graph and the proof rules in Prop. 2.10 and Prop. 3.20 to investigate the fairness properties of the modified ring network.

(g) Repeat (b)–(f) with NoOfBuffers = 3 and with NoOfBuffers = 4 (if possible).

Exercise 3.6.
Consider the philosopher system from Sect. 1.6 and the labelled OS-graphs in Figs. 3.4–3.7.

(a) Compare the OS-graphs of the symmetry specification Φ_{RF} to those of Φ_R. Prove that there must be at least nine philosophers before Φ_{RF} gives us a reduced number of nodes.

(b) Find some properties which you can prove by means of labelled OS-graphs but not by unlabelled OS-graphs.

Chapter 4

Invariants

This chapter deals with place invariants and transition invariants. The basic idea behind place invariants is very similar to that of invariants in program verification. First we formulate some equations – which we postulate to be satisfied independently of the steps that occur. Then we prove that the equations are indeed satisfied, and finally we use them to prove dynamic properties of the modelled system. One difference is the fact that each place invariant of a CP-net is supposed to be satisfied in *all* reachable markings, while a program invariant is usually satisfied only at some distinguished point in the program, e.g., at the start of a loop.

Place invariant analysis has several attractive properties. Firstly, it can be used to verify most of the dynamic properties defined in Chap. 4 of Vol. 1, i.e., reachability, boundedness, home, liveness and fairness properties. Secondly, it is possible to obtain a place invariant for a hierarchical CP-net by composing place invariants of the individual pages. This means it is possible to use place invariants for large systems – without encountering the same kind of complexity problems as we have seen for occurrence graphs. Thirdly, we can use place invariants without fixing system parameters such as the number of sites in a ring network. This means we can prove general system properties independently of the system parameters. Finally, we can construct the place invariants during the design of a system and this will usually lead to an improved design. The main drawback of place invariant analysis is the fact that it requires skills that are higher and more mathematical than those required by many other analysis methods.

Transition invariants are the duals of place invariants. They determine occurrence sequences that have no total effect, i.e., have the same start and end marking.

Section 4.1 contains an informal introduction to place invariants. This is done by means of the data base system and the resource allocation system. Section 4.2 contains the formal definition of place invariants. Here we also define a closely related concept called place flows, and we present a proof rule that allows us to deduce boundedness information from certain kinds of place invariants. Sections 4.3 and 4.4 discuss how place invariants can be found and how this can be supported by computer tools. Finally, Sect. 4.5 investigates transition invariants and Sect. 4.6 deals with invariants for uniform CP-nets.

4.1 Introduction to Place Invariants

The basic idea behind place invariants is to construct equations that are satisfied in all reachable markings. Figure 4.1 shows the data base system from Sect. 1.3 of Vol. 1. The text below the INV-boxes indicate different place invariants, which we shall now construct.

We expect each manager to be either *Inactive*, *Waiting*, or *Performing*. This is expressed by the following equation, which we require to be satisfied for all reachable markings $M \in [M_0\rangle$:

$$M(\text{Inactive}) + M(\text{Waiting}) + M(\text{Performing}) = \text{DBM}.$$

Analogously, we expect each message to be either *Unused*, *Sent*, *Received*, or *Acknowledged* and we expect the system to be either *Active* or *Passive:*

$$M(\text{Unused}) + M(\text{Sent}) + M(\text{Received}) + M(\text{Acknowledged}) = \text{MES}$$

$$M(\text{Active}) + M(\text{Passive}) = 1\text{`e}.$$

These three equations are examples of **place invariants**. Each of them states that a certain set of places has – together – an invariant multi-set of tokens, i.e., a multi-set of tokens which is the same in all reachable markings.

Place invariants may be more complicated than illustrated by the three examples above. It is possible to modify the tokens of some of the involved places before we make the multi-set addition. This is illustrated by the following place invariant, where we use the function *Rec* to map each message (s,r) into the receiver r:

$$M(\text{Inactive}) + M(\text{Waiting}) + \text{Rec}(M(\text{Received})) = \text{DBM}.$$

Without the function *Rec* it would have been impossible to add the three multi-sets, because two of them are over DBM while the last one is over MES. Now let us be a bit more general. Each of the above equations can be written in the form:

$$W_{p_1}(M(p_1)) + W_{p_2}(M(p_2)) + \ldots + W_{p_n}(M(p_n)) = m_{inv}$$

where $p_1, p_2, \ldots, p_n$ are places. Each **weight** W_p is a function mapping from C(p) into some common colour set $A \in \Sigma$ shared by all weights. Finally, m_{inv} is the invariant multi-set. It can be determined by evaluating the left-hand side of the equation in the initial marking (or in any other reachable marking).

As illustrated by the following place invariant there are situations in which we want some of the weights to be negative:

$$M(\text{Performing}) - \text{Rec}(M(\text{Received})) = \emptyset.$$

There are also situations in which we want weights that map some tokens into several elements of type A, instead of just one. This is illustrated by the following place invariant, where the function *Mes* maps each data base manager into $n-1$ tokens of type MES:

$$M(\text{Sent}) + M(\text{Received}) + M(\text{Acknowledged}) - \text{Mes}(M(\text{Waiting})) = \emptyset.$$

To capture the two extensions of weights described above, we introduce **weighted-sets**. A weighted-set is defined in exactly the same way as a multi-set except that we replace $\mathbb{N}$ by $\mathbb{Z}$, i.e., allow the coefficients to be negative. The operations on weighted-sets are similar to the operations on multi-sets. However, for weighted-sets it is always possible to perform subtraction and we can perform scalar-multiplication with negative integers. The set of all weighted-sets over A is denoted by A_{WS}. A formal definition of weighted-sets and their operations will be given in Sect. 4.2.

Having introduced weighted-sets, we require each place $p \in P$ to have a weight which is a function $W_p \in [C(p) \to A_{WS}]$. To apply W_p to a weighted-set $w \in C(p)_{WS}$ (or a multi-set $m \in C(p)_{MS}$), we apply W_p to each individual element of w. This gives us an extended function $W_p{}^* \in [C(p)_{WS} \to A_{WS}]$ defined by:

$$W_p{}^*(w) = \sum_{c \in C(p)} w(c) * W_p(c)$$

for all $w \in C(p)_{WS}$. It is easy to show that $W_p{}^*$ is **linear**, i.e., that it satisfies:

$$W_p{}^*(w_1 + w_2) = W_p{}^*(w_1) + W_p{}^*(w_2)$$

for all weighted-sets $w_1, w_2 \in C(p)_{WS}$. Hence we say that $W_p{}^*$ is the linear extension of W_p. It can be proved that there is a one-to-one correspondence between $[C(p) \to A_{WS}]$ and the linear functions in $[C(p)_{WS} \to A_{WS}]$. Hence we do not need to distinguish between W_p and $W_p{}^*$. A formal definition of linear functions between weighted-sets is given in Sect. 4.2. It is a straightforward generalisation of Def. 3.13.

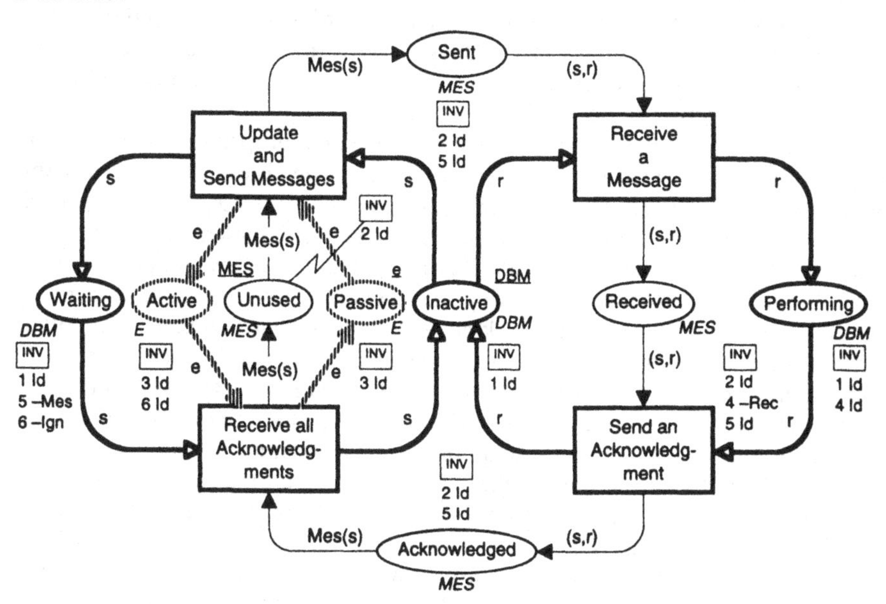

Fig. 4.1. Place invariants for the data base system

The intuition behind a place invariant is the following. For each marking M, we use a set of weights $W = \{W_p\}_{p \in P}$ to calculate a weighted-set called a **weighted sum**:

$$W(M) = \sum_{p \in P} W_p(M(p)).$$

The weighted sum is supposed to be independent of the marking, and hence we require that:

$$(*) \quad W(M) = W(M_0)$$

for all $M \in [M_0\rangle$. As illustrated by our examples above, it is usually the case that many of the weights are **zero functions**, i.e., map each weighted-set into the empty weighted-set, Ø. We shall denote such a function by 0.

Next, let us discuss how we can check that a set of weights really determines a place invariant. Unless the system is trivial, we do not want to calculate W(M) for all reachable markings. To avoid this, we introduce place flows. A set of weights $W = \{W_p\}_{p \in P}$ is a **place flow** iff the following property is satisfied for all binding elements $(t,b) \in BE$:

$$(**) \quad \sum_{p \in P} W_p(E(p,t)\langle b\rangle) = \sum_{p \in P} W_p(E(t,p)\langle b\rangle).$$

The intuition behind a place flow is to require that each binding element (t,b) removes – when the weights are taken into account – a weighted-set of tokens that is identical to the weighted-set of tokens added. For each step $M_1\,[t,b\rangle\, M_2$ that occurs we then know that the weighted sum $W(M_2)$ becomes identical to the weighted sum $W(M_1)$, because the removed tokens are counterbalanced by the added tokens (and because the weights are linear functions). A formal definition of place flows is given in Sect. 4.2.

Note that the invariant property in (*) is a dynamic property, while the flow property in (**) is a static property (i.e., a property which can be checked without considering the set of all reachable markings). Nevertheless, it can be proved that the flow property is sufficient to guarantee the invariant property. The proof is given in Sect. 4.2. In addition to being static, the flow property is also local. It can be checked by considering the individual transitions one at a time. For each transition we need only look at the immediate surroundings. The latter follows from the observation that, in the two sums of (**), we may replace $p \in P$ with $p \in In(t)$ on the left-hand side and with $p \in Out(t)$ on the right-hand side.

We have discussed above how we can use place flows to check whether a set of weights determines a place invariant or not. Now we illustrate how we can use place invariants to prove some of the standard dynamic properties of a CP-net. This is done by considering the data base system with the place invariants shown below (plus many more). For brevity, and improved readability, we omit *M()*. This means we write *Mes(Waiting)* instead of *Mes(M(Waiting))*. We also omit weights that are identity functions and hence we write *Inactive* instead of *Id(M(Inactive))*. The function *Ign* (for ignore) maps each colour (of any type) into $e \in E$. In the last three place invariants, we have $W(M_0) = Ø$, and we have got rid of the negative weights by moving the corresponding terms to the opposite side of the equality sign.

$1:PI_{DBM}$	Inactive+Waiting+Performing = DBM
$2:PI_{MES}$	Unused+Sent+Received+Acknowledged = MES
$3:PI_{E}$	Active+Passive = E
$4:PI_{PER}$	Performing = Rec(Received)
$5:PI_{WA}$	Sent+Received+Acknowledged = Mes(Waiting)
$6:PI_{AC}$	Active = Ign(Waiting)

When we use place invariants to prove dynamic properties of a given CP-net, it is often convenient to show the weights directly on the CP-net – as illustrated in Fig. 4.1. For each place we only list the non-zero weights. For *Waiting* we have three such weights. *Id* belongs to place invariant number 1, *–Mes* belongs to place invariant number 5, and *–Ign* belongs to place invariant number 6.

Let us first show that the place invariants can be used to prove the integer and multi-set bounds postulated in Sect. 4.2 of Vol. 1.

All the multi-set bounds follow from PI_{DBM}, PI_{MES}, and PI_{E}, respectively. The same is true for the integer bounds for *Inactive*, *Unused*, *Passive*, and *Active*. The integer bound for *Waiting* follows from PI_{AC} since we have already shown that $|M(Active)| \le 1$. The integer bound for *Sent*, *Received*, and *Acknowledged* follows from PI_{WA}, since we know that $|M(Waiting)| \le 1$ which implies that $|Mes(M(Waiting))| \le n-1$. The integer bound for *Performing* follows from PI_{PER}, since we know that $|M(Received)| \le n-1$. It is straightforward to construct occurrence sequences showing that all the bounds are optimal, i.e., the smallest possible bounds. We omit this part of the proof.

Next let us prove that the data base system cannot reach a dead marking. The proof is by contradiction. Let us assume that we have a reachable marking M which is dead, i.e., has no enabled transitions. From PI_{DBM} we know that M has n tokens of type DBM, distributed on the places *Inactive*, *Performing*, and *Waiting*. Now let us investigate where these tokens can be positioned.

Let us first assume that at least one data base manager, s, is *Waiting*. From PI_{E} and PI_{AC} it follows that there is exactly one *Waiting* manager and it also follows that the system is *Active*. From PI_{DBM}, we know that the remaining $n-1$ managers are *Inactive* or *Performing*. From PI_{WA} we know that the messages *Mes(s)* are either *Sent*, *Received*, or *Acknowledged*, and we also know that no other messages are in these states. From PI_{PER} it then follows that each manager $r \in DBM-\{s\}$ is *Performing* iff the message (s,r) is *Received* and hence r is *Inactive* iff the message (s,r) is either *Sent* or *Acknowledged*. A message (s,r) on *Sent* would imply that transition *RM* is enabled (since r is *Inactive*). A message (s,r) on *Received* would imply that *SA* is enabled (since r is *Performing*). Hence we conclude that all messages in *Mes(s)* must be *Acknowledged*. However, this implies that *RA* is enabled (since s is *Waiting* and the system is *Active*). Hence we have a contradiction (with the assumption that M is dead).

Next, let us assume that no data base manager is *Waiting*. From the place invariants it is then easy to prove that M is identical to the initial marking. Hence

SM is enabled, and we again have a contradiction (with the assumption that M is dead).

Next, let us prove that M_0 is a home marking. From the proof above, we know that any message on *Sent* can be moved to *Received*, and that any message on *Received* can be moved to *Acknowledged*. This means that, from any marking $M \in [M_0\rangle$, we can reach a marking in which *Sent*, *Received*, and *Performing* have no tokens. We also know that *Waiting* has at most one token. From the place invariants it then follows that we are either already in the initial marking M_0 or in a marking from which M_0 can be reached by an occurrence of *RA*. Hence we have shown that M_0 is a home marking.

Finally, let us prove that SM and RA are strictly live, while RM and SA are live. This follows from the facts that M_0 is a home marking and that it is easy to construct an occurrence sequence which starts in M_0 and contains all binding elements of SM and RA and some binding elements of RM and SA.

As a second example, we consider the resource allocation system with cycle counters, i.e., the CP-net in Fig. 1.7 of Vol. 1. For this net, we have the place invariants shown below (explanation follows).

$1: PI_P$	$PR_1(A+B+C+D+E) = 2`p+3`q$
$2: PI_R$	$R+Q_c(B+C) = 1`e$
$3: PI_S$	$S+Q_c(B)+2*PQ_c(C+D+E) = 3`e$
$4: PI_T$	$T+P_c(D)+(PQ_c+P_c)(E) = 2`e$

PR_1 is the projection function defined in Sect. 4.2 of Vol. 1 and maps each (x,i) colour into x. The function Q_c *counts* the number of q-tokens and maps each (q,i) colour into 1`e and each (p,i) colour into Ø. P_c is defined analogously, and $PQ_c = P_c + Q_c = Ign$. This means we have:

$$PR_1(2`(p,2)+1`(q,1)) = 2`p+1`q$$

$$P_c(2`(p,2)+1`(q,1)) = 2`e$$

$$Q_c(1`(q,1)+3`(q,5)) = 4`e$$

$$PQ_c(2`(p,2)+1`(q,1)) = 3`e.$$

Let us first show that the place invariants can be used to prove the upper bounds postulated in Sect. 4.2 of Vol. 1. The bounds for R, S, and T follow from PI_R, PI_S, and PI_T, respectively. The bounds for A follow from PI_P (and the observation that A can never have p-tokens). The bounds for B follow from PI_P (which tells us that there can be at most two p-tokens), and from PI_R (which tells us that there can be at most one q-token). The bounds for C, D, and E follow from PI_S (which also tells us that at most one of these places has a token). It is straightforward to construct occurrence sequences showing that all the bounds are optimal, and hence we omit this part of the proof. The place invariants can also be used to verify the lower bounds (see Exercise 5.3 of Vol. 1).

Next let us show that the resource allocation system cannot reach a dead marking. The proof is by contradiction: Let us assume that we have a reachable marking M which is dead, i.e., has no enabled transitions. From PI_P we know that M has two p-tokens and three q-tokens, distributed on the places A–E. Now let us investigate in more detail where these tokens can be positioned.

Case a: Assume there are tokens on E. Then T5 is enabled and we have a contradiction (with the assumption that M is dead).

Case b: Assume there are tokens on C and/or D (and no tokens on E). From PI_S it follows that $|M(C)+M(D)| \leq 1$ and then PI_T tells us that there is at least one e-token on T (because $P_c(D) \leq 1`e$ and $(PQ_c+P_c)(E) = \emptyset$). Thus T3 or T4 can occur.

Case c: Assume there are tokens on B (and no tokens on C, D and E). From PI_R it follows that there can be at most one q-token on B and then PI_S tells us that there is at least two e-tokens on S (because $Q_c(B) \leq 1`e$ and $2*PQ_c(C+D+E) = \emptyset$). Thus T2 can occur.

Now we have shown that it is impossible to position the two p-tokens on the places B–E (without violating our assumption of no enabled transition). It is also obvious that the p-tokens cannot be on place A. Thus we conclude that our assumption must be invalid, and hence we have shown that the CP-net cannot reach a dead marking. From the liveness and the cyclic structure of the CP-net, it is possible to prove several other dynamic properties, e.g., that the CP-net has the home space given in Sect. 4.3 of Vol. 1 and is live with respect to the equivalence relation given in Sect. 4.4 of Vol. 1.

In this section we have presented the basic ideas and intuition behind place invariants and place flows. We have also illustrated that place invariants can be used to prove many of the dynamic properties defined in Chap. 4 of Vol. 1. In the following section, we present the formal definition of place invariants and place flows. Moreover, we prove that place flows are sufficient to guarantee place invariants.

4.2 Formal Definition of Place Invariants

In Defs. 4.1–4.2 we define weighted-sets. As explained in Sect. 4.1 a weighted-set is the same as a multi-set except that we allow negative coefficients:

Definition 4.1: A **weighted-set** w, over a non-empty set S, is a function $w \in [S \to \mathbb{Z}]$. The integer $w(s) \in \mathbb{Z}$ is the **number of appearances** of the element s in the weighted-set w.

We usually represent the weighted-set w by a formal sum:

$$\sum_{s \in S} w(s)`s.$$

By S_{WS} we denote the set of all weighted-sets over S. The integers $\{w(s) \mid s \in S\}$ are called the **coefficients** of the weighted-set w, and w(s) is called the **coefficient** of s.

Obviously, there is a one-to-one correspondence between multi-sets and those weighted-sets in which all coefficients are non-negative. Hence we shall consider each multi-set to be a weighted-set. In particular, we use $\emptyset$ to denote the weighted-set in which all coefficients are zero.

Definition 4.2: Addition, subtraction, scalar multiplication, comparison, and **size** of weighted-sets are defined in the following way, for all w, $w_1, w_2 \in S_{WS}$ and all $z \in \mathbb{Z}$:

(i) $w_1 + w_2 = \sum_{s \in S} (w_1(s) + w_2(s))`s$ (addition).

(ii) $w_1 - w_2 = \sum_{s \in S} (w_1(s) - w_2(s))`s$ (subtraction).

(iii) $z * w = \sum_{s \in S} (z * w(s))`s$ (scalar multiplication).

(iv) $w_1 \neq w_2 = \exists s \in S: w_1(s) \neq w_2(s)$ (comparison; $\geq$ and $=$ are
$w_1 \leq w_2 = \forall s \in S: w_1(s) \leq w_2(s)$ defined analogously to $\leq$).

(v) $|w| = \sum_{s \in S} w(s)$ (size).

When $-\infty < |w| < \infty$ we say that w is **finite**.

It should be noted that subtraction is defined for all weighted-sets and that scalar multiplication is defined also for negative integers. When the set S is infinite, the size of a weighted-set $w \in S_{WS}$ may be undefined because the sum is divergent. For a weighted-set $w \in S_{WS}$ we use $-w$ to denote the weighted-set $\emptyset - w$ (i.e., the inverse of w with respect to +). Weighted-sets satisfy a number of nice properties, e.g., those listed in Prop. 2.3 (i)–(viii) of Vol. 1 (replace all $m_i \in S_{MS}$ by $w_i \in S_{WS}$ and all $n_i \in \mathbb{N}$ by $z_i \in \mathbb{Z}$).

Next we define linear functions between weighted-sets (Def. 4.3 and Prop. 4.4 are straightforward generalisations of Def. 3.13 and Prop. 3.14):

Definition 4.3: Let S and R be two sets. A function $F \in [S_{WS} \rightarrow R_{WS}]$ is **linear** iff it satisfies the following property for all $w_1, w_2 \in S_{WS}$:

$$F(w_1 + w_2) = F(w_1) + F(w_2).$$

The set of all linear functions from S_{WS} to R_{WS} is denoted by $[S_{WS} \rightarrow R_{WS}]_L$.

It is easy to verify that a linear function also satisfies the following properties, for all $z \in \mathbb{Z}$ and all $w \in S_{WS}$:

(i) $F(z * w) = z * F(w)$.
(ii) $F(-w) = -F(w)$.
(iii) $F(\emptyset) = \emptyset$.

Proposition 4.4: Each function $F \in [S \to R_{WS}]$ determines a unique linear function $F^* \in [S_{WS} \to R_{WS}]_L$ defined by:

$$F^*(w) = \sum_{s \in S} w(s) * F(s).$$

F^* is called the **linear extension** of F.

Proof: The proof is a straightforward consequence of Defs. 4.2 and 4.3. Hence it is omitted. □

It is easy to prove that each function $F^* \in [S_{WS} \to R_{WS}]_L$ is the linear extension of a function $F \in [S \to R_{WS}]$. Hence there is a one-to-one correspondence between the two sets of functions, and we do not need to distinguish between F and F^*. The following properties of linear functions will be used later in this section and in Sect. 4.4:

Definition 4.5: Let a linear function $F \in [S_{WS} \to R_{WS}]_L$ be given:

(i) F is **non-negative** iff:
$\forall s \in S: F(s) \in R_{MS}$.

(ii) F is **pseudo-surjective** iff:
$\forall r \in R\ \exists w \in S_{WS}\ \exists z \in \mathbb{Z}-\{0\}: F(w) = z`r$.

It is easy to see that F is non-negative iff it always maps multi-sets into multi-sets. It is also easy to see that each surjective function is pseudo-surjective. The same is true for a function of the form $z * F$, where z is a non-zero integer and F is a surjective function. The function *Mes* (from the data base system) is non-negative, but not pseudo-surjective. The functions –Id and $P_C - Q_C$ (from the resource allocation system) are pseudo-surjective, but not non-negative. The function $2 * \text{Id}$ is pseudo-surjective, but not surjective.

Now, we are ready to define place flows and place invariants. To simplify matters, we first present the definitions for a non-hierarchical CP-net. Later we also give the definitions for a hierarchical net.

Definition 4.6: For a non-hierarchical CP-net, a **set of place weights** with range $A \in \Sigma$ is a set of functions $W = \{W_p\}_{p \in P}$ such that $W_p \in [C(p)_{WS} \to A_{WS}]_L$ for all $p \in P$.

(i) W is a **place flow** iff:

$$\forall (t,b) \in BE: \sum_{p \in P} W_p(E(p,t)\langle b\rangle) = \sum_{p \in P} W_p(E(t,p)\langle b\rangle).$$

(ii) W determines a **place invariant** iff:

$$\forall M \in [M_0\rangle: \sum_{p \in P} W_p(M(p)) = \sum_{p \in P} W_p(M_0(p)).$$

In Def. 4.6 we introduce a *set* of weights $\{W_p\}_{p\in P}$. Alternatively, we could have defined a weight *function* W mapping each place p into a weight W(p). We have chosen the first solution because it gives more readable expressions (with fewer parentheses). In Sect. 4.5 we introduce an analogous concept called transition weights. Nevertheless we will often allow ourselves to be a bit sloppy and simply talk about weights (instead of place weights and transition weights). From the context it will then be clear whether we deal with place weights or transition weights.

The following theorem is the heart of place invariant analysis. It tells us that the static property in Def. 4.6 (i) is sufficient and often also necessary to guarantee the dynamic property in Def. 4.6 (ii). Hence it is possible to find place invariants without having to inspect all the reachable markings.

Theorem 4.7: W is a place flow $\Leftrightarrow$ W determines a place invariant.

$\Rightarrow$ is satisfied for all CP-nets.
$\Leftarrow$ is only satisfied when the CP-net does not have dead binding elements.

Proof: Assume that W is a place flow. We first prove that $M_1[Y\rangle M_2$ implies $W(M_1) = W(M_2)$. From the occurrence rule in Def. 2.9 of Vol. 1, we have:

$$\forall p\in P: M_2(p) + \sum_{(t,b)\in Y} E(p,t)\langle b\rangle = M_1(p) + \sum_{(t,b)\in Y} E(t,p)\langle b\rangle,$$

which implies:

$$\sum_{p\in P} W_p\Big(M_2(p) + \sum_{(t,b)\in Y} E(p,t)\langle b\rangle\Big) = \sum_{p\in P} W_p\Big(M_1(p) + \sum_{(t,b)\in Y} E(t,p)\langle b\rangle\Big).$$

From the linearity of the weight functions we get:

$$\sum_{p\in P} W_p(M_2(p)) + \sum_{p\in P}\sum_{(t,b)\in Y} W_p(E(p,t)\langle b\rangle)$$
$$= \sum_{p\in P} W_p(M_1(p)) + \sum_{p\in P}\sum_{(t,b)\in Y} W_p(E(t,p)\langle b\rangle).$$

From the flow property we have:

$$\forall (t,b)\in BE: \sum_{p\in P} W_p(E(p,t)\langle b\rangle) = \sum_{p\in P} W_p(E(t,p)\langle b\rangle),$$

which implies:

$$\sum_{(t,b)\in Y}\sum_{p\in P} W_p(E(p,t)\langle b\rangle) = \sum_{(t,b)\in Y}\sum_{p\in P} W_p(E(t,p)\langle b\rangle),$$

which we can rewrite as:

$$\sum_{p\in P}\sum_{(t,b)\in Y} W_p(E(p,t)\langle b\rangle) = \sum_{p\in P}\sum_{(t,b)\in Y} W_p(E(t,p)\langle b\rangle).$$

The two double sums in this equation are identical to those we had above. Hence we conclude that:

$$\sum_{p \in P} W_p(M_2(p)) = \sum_{p \in P} W_p(M_1(p)),$$

i.e., that $W(M_2) = W(M_1)$.

Next let $M \in [M_0\rangle$ be a reachable marking and let σ be an occurrence sequence that starts in M_0 and ends in M. By applying our result above to each step $M_i\,[Y_i\rangle\, M_{i+1}$ of σ, we conclude that $W(M) = W(M_0)$. Hence we have proved $\Rightarrow$.

Next let us assume that W determines a place invariant and that the CP-net has no dead binding elements. This means each binding element (t,b) has at least one reachable marking M_1 in which it becomes enabled. Let M_2 be the marking determined by $M_1\,[t,b\rangle\, M_2$. By a sequence of arguments similar to those above, it can be seen that $W(M_2) = W(M_1)$ implies that:

$$\sum_{p \in P} W_p(E(p,t)\langle b\rangle) = \sum_{p \in P} W_p(E(t,p)\langle b\rangle)$$

Hence we have proved $\Leftarrow$. □

We are now ready to define place flows and place invariants for hierarchical CP-nets. We then have a weight for each place instance group. This means all place instances belonging to a place instance group automatically get the same weight (and together they make *one* contribution to the weighted sum). For a place instance group $p'' \in PIG$ we use $C(p'')$ to denote the common colour set. It should be noted that the flow property in Def. 4.8 (i) is defined in terms of place instances, while the invariant property in Def. 4.8 (ii) is defined in terms of place instance groups.

Definition 4.8: For a hierarchical CP-net, a **set of place weights** with range $A \in \Sigma$ is a set of functions $W = \{W_{p''}\}_{p'' \in PIG}$ such that $W_{p''} \in [C(p'')_{WS} \to A_{WS}]_L$ for all $p'' \in PIG$. For a place instance $p' \in PI$ we use $W_{p'}$ to denote the weight of the place instance group $p'' \in PIG$ to which p' belongs.

(i) W is a **place flow** iff:

$$\forall (t',b) \in BE: \sum_{p' \in PI} W_{p'}(E(p',t')\langle b\rangle) = \sum_{p' \in PI} W_{p'}(E(t',p')\langle b\rangle).$$

(ii) W determines a **place invariant** iff:

$$\forall M \in [M_0\rangle: \sum_{p'' \in PIG} W_{p''}(M(p'')) = \sum_{p'' \in PIG} W_{p''}(M_0(p'')).$$

Theorem 4.7 remains valid when we go from non-hierarchical CP-nets to hierarchical nets (i.e., replace Def. 4.6 by Def. 4.8). The basic idea behind the proof is the same, but the technical details are more cumbersome. Hence we shall omit the proof.

For different kinds of occurrence graphs, Chaps. 1–3 present a large number of proof rules. They make it easy for the user to deduce net properties from occurrence graph properties. Unfortunately, the situation is less favourable for place invariants. Except for the boundedness properties there are no canned rules to prove the different kinds of dynamic properties. Instead the user has to perform a sequence of arguments that is more like a mathematical proof. This was illustrated by our examples in Sect. 4.1.

Below we present a proof rule for the boundedness properties. A place flow is said to be non-negative iff all the place weights are non-negative functions; see Def. 4.5 (i).

Proposition 4.9: Let $p' \in PI$ be a place instance and $W = \{W_{p''}\}_{p'' \in PIG}$ a non-negative place flow with $W_{p'} = Id$. For the **boundedness properties** we then have the following proof rules:

(i) $W(M_0)$ is an upper multi-set bound for p'.
(ii) $W(M_0)$ finite $\Rightarrow$ $|W(M_0)|$ is as an upper integer bound for p'.

Proof: W determines an invariant that can be written in the form:

$$M(p') + \sum_{p'' \in PIG - \{[p']\}} W_{p''}(M(p'')) = W(M_0)$$

where [p'] denotes the page instance group to which p' belongs. Since W is non-negative, we know that the sum over $PIG - \{[p']\}$ evaluates to a multi-set (and not just a weighted-set). Hence we conclude that:

$$M(p') \leq W(M_0)$$

where $\leq$ denotes comparison of multi-sets. This finishes the proof of (i). Property (ii) is a direct consequence of (i). □

The bounds in Prop. 4.9 are not guaranteed to be optimal, i.e., the smallest possible bounds. However, in practice, we often obtain bounds that are optimal, or nearly optimal. For example, consider the data base system, and the place invariants PI_{DBM}, PI_{MES} and PI_E. With respect to multi-set bounds we obtain all the optimal bounds postulated in Sect. 4.2 of Vol. 1. With respect to integer-bounds the situation is less favourable. It is only for *Inactive*, *Unused*, *Passive*, and *Active* that we obtain the optimal integer bounds.

In the next two sections we discuss how to construct place invariants, i.e., how to find sets of place weights satisfying the flow properties in Defs. 4.6 (i) and 4.8 (i). There are two alternative approaches.

The first approach involves a fully automatic calculation of all place flows for the given CP-net. As we shall see, it is indeed possible to make such a calculation. This is done by representing the CP-net as a matrix and solving a homogeneous matrix equation. Usually there is an infinite number of solutions to the matrix equation and hence an infinite number of place flows. It is possible to represent the solutions by a finite set of flows, from which the others can be generated. However, from such a set, it is often not at all easy to find the place invariants that are useful to prove dynamic properties of the system considered. The automatic approach is discussed in Sect. 4.3.

The second approach is interactive. It builds on the idea that the modeller already has a considerable amount of knowledge about the system and its properties. Hence it is not difficult for the modeller to find some sets of place weights which are expected to determine place invariants. The expected place invariants can be automatically verified and they can be immediately interpreted by the modeller – since they have been formulated by himself. The interactive approach is discussed in Sect. 4.4.

4.3 Automatic Calculation of Place Invariants

In this section we discuss how to make an automatic calculation of place invariants. To simplify matters we first consider a non-hierarchical CP-net. At the end of the section we discuss the modifications needed to cope with hierarchical nets.

First we introduce the **incidence matrix**. This is another way to represent a CP-net. The alternative representation turns out to be convenient when we deal with calculation of invariants. The incidence matrix of a non-hierarchical CP-net contains a row for each place $p \in P$ and a column for each transition $t \in T$. We assume that the elements of P and T are ordered in some arbitrary (but fixed) way. Each matrix element I(p,t) is a linear function in $[B(t)_{WS} \rightarrow C(p)_{WS}]_L$. It is defined by:

$$I(p,t)(b) = E(t,p)\langle b \rangle - E(p,t)\langle b \rangle.$$

This means I(p,t)(b) describes how the marking of the place p is changed when the transition t occurs with the binding b.

As an example, let us consider the incidence matrix for the data base system. The incidence matrix is shown below. For simplicity, we have assumed that transitions RM and SA have the guard $[s \neq r]$. In this way we avoid bindings that never can occur. We use the same functions as in Sect. 4.1. We also use identity functions (denoted by Id) and zero functions (denoted by an empty matrix entrance). To improve readability we list the places (and their colour sets) at the left border of the incidence matrix and the transitions (and their sets of bindings) at the top border. However, strictly speaking, it is only the 9 times 4 elements in the lower right part that constitute the incidence matrix.

Incidence Matrix for Data Base System		SM	RA	RM	SA
		DBM	DBM	MES	MES
Inactive	DBM	–Id	Id	–Rec	Rec
Waiting	DBM	Id	–Id		
Performing	DBM			Rec	–Rec
Unused	MES	–Mes	Mes		
Sent	MES	Mes		–Id	
Received	MES			Id	–Id
Acknowledged	MES		–Mes		Id
Passive	E	–Ign	Ign		
Active	E	Ign	–Ign		

The incidence matrix can be automatically derived from the other representations of CP-nets (such as the many-tuples in Defs. 2.5 and 3.1 of Vol. 1 and the

CPN diagrams used by the CPN editor). It should be noted that a little bit of information is lost, because the functions of the incidence matrix only allow us to determine the weighted-set:

E(t,p)<b> – E(p,t)<b>

and not the two multi-sets E(t,p)<b> and E(p,t)<b>, separately. For the telephone system in Sect. 3.2 of Vol. 1, this implies that the matrix element for the place *Connection* and the transition *RepConRec* becomes a zero function although there are arcs between the two nodes. Below, we illustrate the usefulness of the matrix representation for calculating place invariants. But first we discuss how also markings and steps can be represented as matrices.

We represent each marking $M \in \mathbb{M}$ as a **column vector**, i.e., a matrix with a single column (and a row for each place). For each place $p \in P$ we have the matrix element $M(p) \in C(p)_{MS}$. Analogously, we represent each step Y as a column vector (with a row for each transition). For each transition $t \in T$ we have the matrix element $Y(t) \in B(t)_{MS}$. As examples, we show the representation of the markings M_2 and M_3 and the step Y_{2b} from Sect. 2.3 of Vol. 1. An empty entry in the vector indicates that the corresponding matrix element is the empty multi-set. As with the incidence matrix, we improve readability by listing the places/transitions to the left of the column vectors:

Marking M_2	
Inac	$1`d_2+1`d_3+1`d_4$
Wait	$1`d_1$
Perf	
Un	$MES-Mes(d_1)$
Sent	$1`(d_1,d_2)+1`(d_1,d_3)+1`(d_1,d_4)$
Rec	
Ack	
Pas	
Ac	$1`e$

Marking M_3	
Inac	$1`d_3$
Wait	$1`d_1$
Perf	$1`d_2+1`d_4$
Un	$MES-Mes(d_1)$
Sent	$1`(d_1,d_3)$
Rec	$1`(d_1,d_2)+1`(d_1,d_4)$
Ack	
Pas	
Ac	$1`e$

Step Y_{2b}	
SM	
RA	
RM	$1`(d_1,d_2)+1`(d_1,d_4)$
SA	

Next we show how the matrices above can be used to describe the effect of a step. For standard matrices (with numbers as elements) we define the sum of two matrices A and B (with n rows and m columns) to be the matrix $A+B$ defined by:

$$(A+B)_{ij} = A_{ij}+B_{ij}$$

for all $i \in 1..n$ and all $j \in 1..m$. We define the product of a matrix A (with n rows and m columns) and a matrix B (with m rows and p columns) to be the matrix $A * B$ (with n rows and p columns) defined by:

$$(A * B)_{ik} = \sum_{j=1}^{m} A_{ij} * B_{jk}.$$

for all $i \in 1..n$ and all $k \in 1..p$. The operations in the right-hand side of the two equations above represent addition and multiplication of numbers. It is known from mathematics that the above definitions of matrix operations can also be applied when the matrix elements are no longer numbers. To permit this, the elements of A and B must have two laws of composition allowing us to perform the necessary "additions" and "multiplications". The addition law must be associative and commutative, while the multiplication law must be associative and distributive (with respect to addition). With these assumptions we can show that matrix products and matrix additions are well defined and have the usual properties, e.g., that matrix addition is associative and commutative, while matrix multiplication is associative and distributive (with respect to matrix addition). The proofs of these observations are straightforward. Hence they are omitted.

Now let us return to the incidence matrix and the column vectors for markings and steps. The elements of the incidence matrix are functions (mapping bindings into weighted-sets of token colours). The elements of the column vectors are multi-sets (of token colours and bindings, respectively).

We wish to multiply the incidence matrix I by a step Y, i.e., to define the matrix product $I * Y$. To do this, we need to make a multiplication of the form $I(p,t) * Y(t)$ where $I(p,t)$ is a function belonging to $[B(t)_{WS} \rightarrow C(p)_{WS}]_L$ while $Y(t)$ is a multi-set over $B(t)$. A rather straightforward (and successful) choice is to define $I(p,t) * Y(t)$ to be the result of applying the function $I(p,t)$ to the multi-set $Y(t)$. This works since $Y(t)$ belongs to $B(t)_{MS} \subseteq B(t)_{WS}$. We also need to add two products $I(p,t') * Y(t')$ and $I(p,t'') * Y(t'')$ to each other (in order to perform the summation operation). However, this is straightforward. We simply use the standard addition operation for weighted-sets; see Def. 4.2 (i). With the observations above, we define the product $I * Y$ to be:

$$(I * Y)(p) = \sum_{t \in T} I(p,t)(Y(t))$$

for all $p \in P$. The result is a column vector with a row for each place. The matrix element $(I * Y)(p)$ describes how the marking of the place p is changed when the step Y occurs. This can be seen from the following calculation:

$$\begin{aligned}(I * Y)(p) &= \sum_{t \in T} I(p,t)(Y(t)) \\ &= \sum_{t \in T} \sum_{b \in Y(t)} (E(t,p)<b> - E(p,t)<b>) \\ &= \sum_{(t,b) \in Y} E(t,p)<b> - \sum_{(t,b) \in Y} E(p,t)<b>.\end{aligned}$$

The two final sums are identical to the added and to the removed tokens respectively in the occurrence rule in Def. 2.9 of Vol. 1. Hence we conclude that for an occurrence $M_1\,[Y\rangle\, M_2$ we can calculate the marking $M_2(p)$ by means of the following equation:

$$M_2(p) = M_1(p) + (I * Y)(p).$$

Thus we have:

$$M_2 = M_1 + I * Y$$

where we define addition of the two matrices M_1 and $I * Y$ to be element wise addition of weighted-sets. This works because $M_1(p)$ and $(I * Y)(p)$ both belong to $C(p)_{WS}$.

Now let us return to the column vectors M_2, M_3, and Y_{2b} defined above. From the discussion in Sect. 2.3 of Vol. 1, we know that $M_2\,[Y_{2b}\rangle\, M_3$, hence:

$$M_3 = M_2 + I * Y_{2b},$$

i.e.:

$$\begin{array}{|c|}\hline 1\text{`}d_3 \\ 1\text{`}d_1 \\ 1\text{`}d_2 + 1\text{`}d_4 \\ \hline MES - Mes(d_1) \\ 1\text{`}(d_1,d_3) \\ 1\text{`}(d_1,d_2) + 1\text{`}(d_1,d_4) \\ \\ \hline \\ 1\text{`}e \\ \hline \end{array} = \begin{array}{|c|}\hline 1\text{`}d_2 + 1\text{`}d_3 + 1\text{`}d_4 \\ 1\text{`}d_1 \\ \\ \hline MES - Mes(d_1) \\ 1\text{`}(d_1,d_2) + 1\text{`}(d_1,d_3) + 1\text{`}(d_1,d_4) \\ \\ \\ \hline \\ 1\text{`}e \\ \hline \end{array} +$$

$$\begin{array}{|cc|cc|}\hline -Id & Id & -Rec & Rec \\ Id & -Id & & \\ & & Rec & -Rec \\ \hline -Mes & Mes & & \\ Mes & & -Id & \\ & & Id & -Id \\ & -Mes & & Id \\ \hline -Ign & Ign & & \\ Ign & -Ign & & \\ \hline \end{array} * \begin{array}{|c|}\hline \\ \\ \hline 1\text{`}(d_1,d_2) + 1\text{`}(d_1,d_4) \\ \\ \hline \end{array}$$

The reader is encouraged to check that the above matrix equation is correct. Next let us consider a set of place weights, i.e., a set of functions $W = \{W_p\}_{p \in P}$ such that $W_p \in [C(p)_{WS} \rightarrow A_{WS}]_L$ for all $p \in P$. It is obvious that W can be represented as a matrix. For reasons which will be clear in a moment, it turns out to be convenient to represent it as a **row vector**, i.e., a matrix with a single row (and a column for each place). For each $p \in P$ we have the matrix element W_p. Now let us again look at an example. In Sect. 4.1 we found 6 place flows for the data base system. The fifth of these is represented as follows:

	Inac	Wa	Perf	Un	Sent	Rec	Ack	Pas	Ac
PI_{WA}		$-$Mes			Id	Id	Id		

Having defined a matrix representation for a set of place weights $W = \{W_p\}_{p \in P}$, we can now consider the matrix product $W * I$. To make this well defined, we need a "multiplication" operation, which works on a pair, where the first element is a function $W_p \in [C(p)_{WS} \rightarrow A_{WS}]_L$ and the second element is a function $I(p,t) \in [B(t)_{WS} \rightarrow C(p)_{WS}]$. A rather straightforward (and successful) choice is to define $W_p * I(p,t)$ to be the functional composition of the two functions W_p and $I(p,t)$. This works since the domain of W_p is identical to the range of $I(p,t)$. We also need to be able to add two products $W_{p'} * I(p',t)$ and $W_{p''} * I(p'',t)$ to each other (in order to perform the summation operation). However, this is straightforward. We simply use the standard addition operation for functions, i.e., $(F+G)(x) = F(x) + G(x)$. With the observations above, we define the product $W * I$ to be:

$$(W * I)(t) = \sum_{p \in P} W_p \circ I(p,t)$$

for all $t \in T$. The result is a row vector with a column for each transition. The matrix element in $(W * I)(t)$ describes how the weighted sum $W(M)$ is changed when the transition t occurs. This follows from the following sequence of calculations in which we use the linearity of the involved functions:

$$\begin{aligned} ((W * I)(t))(b) &= \sum_{p \in P} (W_p \circ I(p,t))(b) \\ &= \sum_{p \in P} W_p(E(t,p)<b> - E(p,t)<b>) \\ &= \sum_{p \in P} W_p(E(t,p)<b>) - \sum_{p \in P} W_p(E(p,t)<b>). \end{aligned}$$

From this it is easy to see that $W = \{W_p\}_{p \in P}$ is a place flow iff $W * I$ is the zero matrix, i.e., the matrix where all elements are zero functions. Thus we can find place flows and thus place invariants by solving the homogeneous matrix equation:

$$W * I = 0$$

where the row vector W is the unknown.

Finally, we consider matrix products of the form $W * M$, where we define the product $W_p * M(p)$ to mean application of the function $W_p \in [C(p)_{WS} \rightarrow A_{WS}]_L$ to the multi-set $M(p) \in C(p)_{MS} \subseteq C(p)_{WS}$ and where addition is the standard addition for weighted-sets. The result is a matrix with one row and one column, i.e., a single element in A_{WS}:

$$W * M = \sum_{p \in P} W_p(M(p)).$$

From this we conclude that the matrix product $W * M$ is identical to the weighted sum $W(M)$ defined in Sect. 4.2.

Now we can give a much briefer proof of ⇒ in Theorem 4.7. We assume that W is a place flow, i.e., that $W * I = 0$ and we consider an occurrence $M_1 [Y\rangle M_2$. Then we have:

$$\begin{aligned}
W(M_2) &= W * M_2 && \text{(definition of } W * M_2\text{)}\\
&= W * (M_1 + I * Y) && (M_2 = M_1 + I * Y)\\
&= W * M_1 + W * (I * Y) && \text{(distributivity of * over +)}\\
&= W(M_1) + (W * I) * Y && \text{(associativity of *)}\\
&= W(M_1) && (W * I = 0).
\end{aligned}$$

The idea behind the "new" proof is identical to that of Theorem 4.7. However, we have now developed a notation which makes it possible to formulate a more succinct representation of the manipulations.

Next let us consider how we can solve the matrix equation $W * I = 0$. For matrices where the elements are numbers, it is known that the solutions can be determined by a method called Gaussian elimination. A description of this method can be found in almost any introductory book on linear equation systems. However, with our more general kind of matrices there are a couple of problems.

The first problem is the fact that we may have matrix elements for which the inverse (with respect to $\circ$) does not exist. Intuitively this means we cannot always divide matrix elements by each other. This problem can be circumvented, but the algorithms become rather complex.

The second problem is that it may be difficult to interpret our solutions to the matrix equation. As long as we deal with numbers we can make complex calculations such as:

$$x + (y - z / v) / w$$

and we still have a number which we can understand immediately (provided that v and w are non-zero). However, when x, y, z, v, and w are functions, it is not so easy to understand what the function $x + (y - z / v) / w$ does – even though it may be well defined.

The third problem is that the matrix equation usually has infinitely many solutions. It is often possible to represent the solutions by a finite set of flows – from which the others can be generated. However, from such a set, it is usually not at all easy to find those place invariants which are useful to prove dynamic properties of the considered system. Hence the user is left with the problem of determining the useful place flows, and this problem is often as difficult as finding the generating set of solutions.

There are many different techniques for automatic calculation of place flows. A detailed description of these is outside the scope of this book. References to some of the more important methods are given in the bibliographical remarks. Most of the methods are rather complex and draw heavily on results from linear algebra and linear programming.

To deal with a hierarchical CP-net we may construct an incidence matrix which has a row for each place instance group $p'' \in PIG$ and a column for each transition instance $t' \in TI$. Each matrix element, $I(p'',t')$ is a linear function in $[B(t')_{WS} \rightarrow C(p'')_{WS}]_L$ where $B(t')$ is the set of bindings for the transition of which t' is an instance and $C(p'')$ is the common colour set of the places that contribute to p''. The matrix element is defined by:

$$I(p'',t')(b) = \sum_{p' \in p''} E(t',p')\langle b\rangle - \sum_{p' \in p''} E(p',t')\langle b\rangle.$$

This means $I(p'',t')(b)$ describes how the marking of the place instance group p'' is changed when the transition instance t' occurs with the binding b.

With the above definition it can be proved that the incidence matrix of a hierarchical CP-net is identical to the incidence matrix of the CP-net that is the non-hierarchical equivalent (see Def. 3.7 of Vol. 1). This means we can calculate place invariants in a similar way to that described above. The only difference is that the incidence matrix of a hierarchical CP-net will often be quite large, and hence we may easily run into complexity problems. To avoid such problems, we may instead define an incidence matrix for each page. This is done in exactly the same way as for a non-hierarchical CP-net, and hence we have a row for each place (belonging to the page) and a column for each transition (belonging to the page). We now have a set of homogeneous matrix equations, one for each page. To calculate the place flows of the hierarchical CP-net, we must solve all the matrix equations and combine the solutions in such a way that they become consistent with each other. This means all place instances belonging to a place instance group must have identical weights.

4.4 Interactive Calculation of Place Invariants

The second approach to calculation of place invariants does not start from scratch. Instead it builds on the idea that the modeller has already a considerable amount of knowledge about the system and its properties. Hence it is not difficult for the modeller to find some sets of place weights that are expected to determine place invariants. The potential place invariants may be derived, e.g., from the system specification and from the modeller's knowledge of the expected system properties. The place invariants may be specified during the analysis of the CP-net. However, it is much more useful (and also easier) to specify the place invariants while the CP-net is being created. This means we construct the place invariants as an integrated part of the model design (rather as a good programmer specifies a loop invariant at the moment he creates a loop). In programming, it would be most unusual (and unrealistic) first to write a large program (without even thinking about invariants) and then expect an automatic construction of useful invariants that can be easily interpreted by the programmer.

In the previous paragraph we have advocated that the modeller should be responsible for the construction of potential place invariants. However, there are several ways to support the process by computer tools. Hence it is better to think about the construction as an interactive process than to think of it as purely manual.

Check of place flows

It is possible to automate the check of the flow properties in Defs. 4.6 (i) and 4.8 (i). For each transition t, we need to show that the function F_t that maps each binding $b \in B(t)$ into the weighted-set:

$$F_t(b) = \sum_{p \in P} W_p(E(t,p)\langle b\rangle) - \sum_{p \in P} W_p(E(p,t)\langle b\rangle)$$

is the zero function, i.e., that F_t maps all bindings of t into Ø. When the set of bindings is small we can simply test each binding in turn. However, when t has a large set of bindings, or even an infinite set, we need a more subtle method. At the end of this section we describe a place invariant tool that is integrated with the CPN tools described in Chap. 6 of Vol. 1. The place invariant tool verifies the flow property by representing the function F_t as a lambda expression. This is possible because the arc expressions, guards, and place weights are specified in the functional programming language Standard ML, which compiles into lambda expressions. By means of lambda reductions it is then checked whether F_t is the zero function or not. The check is fully automatic, and in case of errors it indicates exactly where the problems are by identifying the binding elements that violate the flow property. Hence it is usually rather straightforward to figure out how to modify the CP-net or the place weights so that the flow property becomes valid. Arc expressions, guards, and place weights may be arbitrarily complex, and hence there exist transitions for which the reduction is impossible (or takes a long time). However, the place invariant tool is able to cope with most of the arc expressions, guards, and place weights found in "standard" CP-nets.

Reduction of incidence matrix

It is often possible to make a reduction of the incidence matrix. This is done by using a set of transformation rules which simplify the matrix without changing the set of place flows. The transformation rules are inspired by the manipulations that we use when we solve a homogeneous matrix equation by means of Gaussian elimination. It should be noted that standard Gaussian elimination solves an equation $I * W = 0$, while we want to solve the equation $W * I = 0$. This means our manipulations deal with columns instead of rows.

As we perform our transformations there will no longer be a column for each transition and a row for each place. Instead we will have a set of columns T^* and a set of rows P^*. Each column $t^* \in T^*$ will have a non-empty set $B(t^*)$ such that $B(t^*)_{WS}$ is the common domain for all the matrix elements in the column t^*. The column will represent a transition or a linear combination of several transitions. Analogously, each row $p^* \in P^*$ will have a non-empty set $C(p^*)$ such that $C(p^*)_{WS}$ is the common range for all the matrix elements in the row

p^*. The row p^* will represent one or more places and each of these, p, will have a **weight-factor** $WF_p \in [C(p)_{WS} \to C(p^*)_{WS}]_L$. The weight-factor tells us how to calculate a weight W_p, for the place p, from a weight $W_{p^*} \in [C(p^*)_{WS} \to A_{WS}]_L$ determined for the row p^*. This is done as follows:

$$W_p = W_{p^*} \circ WF_p.$$

The result is a function in $[C(p)_{WS} \to A_{WS}]_L$ – as required. The places represented by a row are also said to be **attached** to the row. When we show a matrix, we list each weight-factor in parentheses immediately after the corresponding place. In the examples of this book all weight factors are Id and hence they are omitted. In the original incidence matrix each place is attached to a row of its own with weight-factor Id. Analogously, each transition has a column of its own.

A column $t^* \in T^*$ is **linearly dependent** on a set of other columns $T^+ \subseteq T^* - \{t\}$ iff there exist a colour set $A \in \Sigma$ and a function $F_t \in [A_{WS} \to B(t)_{WS}]_L$ for all $t \in T^+ \cup \{t^*\}$ such that F_{t^*} is pseudo-surjective and:

$$\sum_{t \in T^+ \cup \{t^*\}} I(p,t) \circ F_t = 0$$

for all $p \in P^*$. It should be noted that it is only F_{t^*} which we require to be pseudo-surjective. The other functions only have to be linear. It should also be noted that although t^* is linearly dependent on T^+, we do not always know that each of the other columns $t \in T^+$ is linearly dependent on $(T^+ \cup \{t^*\}) - \{t\}$. Hence there are some standard concepts from linear algebra that we cannot define, e.g., basis and rank.

We are now ready to present the transformation rules. The soundness of the rules will be established via Theorem 4.10. Rule 1 is similar to the Gaussian elimination rule that allows us to multiply all the elements in a row with a common non-zero integer. Rule 2 is similar to the Gaussian elimination rule that allows us to add an arbitrary multiple of one row to another row.

Rule 1: For a column $t \in T^*$, a colour set $A \in \Sigma$, and a pseudo-surjective function $F \in [A_{WS} \to B(t)_{WS}]_L$, we may replace column t by a new column t^* where $B(t^*) = A$ while:

$$I(p,t^*) = I(p,t) \circ F$$

for all $p \in P^*$.

Rule 2: For two columns $t', t'' \in T^*$ and a function $F \in [B(t')_{MS} \to B(t'')_{WS}]_L$, we may replace column t' by a new column t^* where $B(t^*) = B(t')$ while:

$$I(p,t^*) = I(p,t') + I(p,t'') \circ F$$

for all $p \in P^*$.

Rule 3: A column in which all elements are zero functions may be removed. The same is true for a column which is linearly dependent on a set of other columns.

Rule 4: Assume we have a column t' where all elements are zero functions except I(p',t'), which is pseudo-surjective. We may then delete column t' and row p'.

Rule 5: Assume we have a column t' where all elements are zero functions except I(p',t') and I(p",t'), which are pseudo-surjective and each other's negation. We may then delete column t' and replace row p' and row p" by a new row p* where C(p*) = C(p') = C(p") and:

$$I(p^*,t) = I(p',t) + I(p'',t)$$

for all $t \in T^*$. The places attached to p' or p" are instead attached to p*. They get the same weight-factor as before the transformation.

Rule 6: Assume we have a column t' where all elements are zero functions except I(p',t') and I(p",t'), which are pseudo-surjective and identical. We may then delete column t' and replace row p' and row p" by a new row p* where C(p*) = C(p') = C(p") and:

$$I(p^*,t) = I(p',t) - I(p'',t)$$

for all $t \in T^*$. The places attached to p' or p" are instead attached to p*. Those that were attached to p' get the same weight-factor as before the transformation. Those attached to p" get a weight-factor which is the negation of the one they had before the transformation.

Rule 7: Assume we have a column t' where all elements are zero functions except I(p",t'), which is pseudo-surjective, and I(p',t'), which can be written as:

$$I(p',t') = -F \circ I(p'',t')$$

where F is a linear function in $[C(p'')_{WS} \rightarrow C(p')_{WS}]_L$. We may then delete column t' and replace row p' and row p" by a new row p* where C(p*) = C(p') and:

$$I(p^*,t) = I(p',t) + F \circ I(p'',t)$$

for all $t \in T^*$. The places attached to p' or p" are instead attached to p*. The places that were attached to p' get the same weight-factor as before the transformation. Those attached to p" get the weight-factor $F \circ WF_p$ where WF_p is the weight factor before the transformation. If $F \circ WF_p$ is the zero function, we discard the place (instead of attaching it to p*). □

With a little bit of work, we can see that rules 4–6 are special cases of rule 7. They cover the situations in which F is 0, Id, and –Id, respectively. It would be sufficient to have rule 7. However, we include rules 4–6, because they are much easier to understand and cover the majority of the situations in which rule 7 is used.

Let P* be the rows of a matrix R obtained from an incidence matrix I by means of rules 1–7. A set of weights $\{W_p\}_{p \in P}$ is **determined** by the matrix R iff the matrix equation $U * R = 0$ has a solution $U = \{U_{p^*}\}_{p^* \in P^*}$ such that:

$$W_p = \begin{cases} U_{p^*} \circ WF_p & \text{iff p is attached to row } p^* \text{ with weight-factor } WF_p \\ 0 & \text{iff p is not attached to any row in R.} \end{cases}$$

The following theorem shows that the transformation rules are sound, in the sense that the reduced matrix determines exactly those sets of weights which are place flows for the incidence matrix. The proof is a bit complicated and it may be skipped by readers who are primarily interested in the practical application of the transformation rules.

Theorem 4.10: The sets of weights determined by a reduced matrix R is identical to the set of place flows for the original matrix I.

Proof: The proof is by structural induction (over the number of transformations). Each application of a rule transforms a matrix I' into another matrix I". We have to show that the two matrices determine the same sets of weights. All summations are over the rows of I' and I", respectively. We use p to denote such a row.

Rule 1 replaces column t by column t* but leaves all other columns and all weight-factors unaltered. Hence we have to prove that:

$$\Sigma\ U_p \circ I'(p,t) = 0 \ \Leftrightarrow\ \Sigma\ U_p \circ I''(p,t^*) = 0,$$

i.e., that:

$$\Sigma\ U_p \circ I'(p,t) = 0 \ \Leftrightarrow\ \Sigma\ U_p \circ (I'(p,t) \circ F) = 0.$$

By linearity of the involved functions we get:

$$(\Sigma\ U_p \circ (I'(p,t) \circ F))(a) = (\Sigma\ U_p \circ I'(p,t))(F(a))$$

for all $a \in A$. Hence it is sufficient to prove that:

$$\forall w \in B(t)_{WS}: (\Sigma\ U_p \circ I'(p,t))(w) = \emptyset$$
$$\Leftrightarrow\ \forall a \in A_{WS}: (\Sigma\ U_p \circ I'(p,t))(F(a)) = \emptyset.$$

$\Rightarrow$ is trivial since each F(a) belongs to $B(t)_{WS}$. $\Leftarrow$ follows from the pseudo-surjectivity of F and the linearity of the involved functions.

Rule 2 replaces column t' by column t* but leaves all other columns and all weight-factors unaltered. Hence it is sufficient to prove that:

$$\Sigma\ U_p \circ I'(p,t') = 0 \ \wedge\ \Sigma\ U_p \circ I'(p,t'') = 0$$
$$\Leftrightarrow\ \Sigma\ U_p \circ I''(p,t^*) = 0 \ \wedge\ \Sigma\ U_p \circ I''(p,t'') = 0,$$

i.e., that:

$$\Sigma\ U_p \circ I'(p,t') = 0 \ \wedge\ \Sigma\ U_p \circ I'(p,t'') = 0$$
$$\Leftrightarrow\ \Sigma\ U_p \circ (I'(p,t') + I'(p,t'') \circ F) = 0 \ \wedge\ \Sigma\ U_p \circ I'(p,t'') = 0.$$

This is a direct consequence of the linearity of the involved functions.

Rule 3. It is trivial that we can remove zero columns without changing the sets of weights determined by a matrix. A column that is linearly dependent on a set of other columns can be turned into a zero column by using rule 2 a number of times.

Rule 4. The column t' tells us that:

$$\Sigma\ U_p \circ I'(p,t') = U_{p'} \circ I'(p',t') = 0.$$

From the pseudo-surjectivity of I'(p',t') and the linearity of $U_{p'}$ we then conclude that $U_{p'} = 0$. From this it is easy to see that we can remove column t' and row p' without changing the sets of weights determined by the matrix.

Rule 5. The column t' tells us that:

$$\Sigma\ U_p \circ I'(p,t') = U_{p'} \circ I'(p',t') + U_{p''} \circ I'(p'',t')$$
$$= U_{p'} \circ I'(p',t') - U_{p''} \circ I'(p',t') = (U_{p'} - U_{p''}) \circ I'(p',t') = 0.$$

From the pseudo-surjectivity of I'(p',t') and the linearity of $U_{p'}$ and $U_{p''}$ we then conclude that $U_{p''} = U_{p'}$. From this it is easy to see that we can remove column t' and add the rows of p' and p" without changing the sets of weights determined by the matrix.

Rule 6. Similar to the proof of rule 5. This time we conclude that $U_{p''} = -U_{p'}$.

Rule 7. Similar to the proof of rule 5. This time we conclude that $U_{p''} = U_{p'} \circ F$.

□

The transformation rules may seem a bit primitive. However, they are quite powerful – for our purpose. As an example, let us consider how they can be used to reduce the incidence matrix for the data base system. As we have seen in Sect. 4.3, the incidence matrix looks as follows:

Incidence Matrix for Data Base System		SM	RA	RM	SA
		DBM	DBM	MES	MES
Inactive	DBM	–Id	Id	–Rec	Rec
Waiting	DBM	Id	–Id		
Performing	DBM			Rec	–Rec
Unused	MES	–Mes	Mes		
Sent	MES	Mes		–Id	
Received	MES			Id	–Id
Acknowledged	MES		–Mes		Id
Passive	E	–Ign	Ign		
Active	E	Ign	–Ign		

By means of rule 2, we can add the first column to the second column, and we can add the third column to the fourth. This gives us the following matrix:

Reduced Matrix for Data Base System		T_1 DBM	T_2 DBM	T_3 MES	T_4 MES
Inactive	DBM	–Id		–Rec	
Waiting	DBM	Id			
Performing	DBM			Rec	
Unused	MES	–Mes			
Sent	MES	Mes	Mes	–Id	–Id
Received	MES			Id	
Acknowledged	MES		–Mes		Id
Passive	E	–Ign			
Active	E	Ign			

By means of rule 3 we can then remove T_2, because it is linearly dependent on T_4. By means of rule 5 we add the rows for *Sent* and *Acknowledged* and remove T_4. This gives us the following matrix:

Reduced Matrix for Data Base System		T_1 DBM	T_3 MES
Inactive	DBM	–Id	–Rec
Waiting	DBM	Id	
Performing	DBM		Rec
Unused	MES	–Mes	
Sent, Ack.	MES	Mes	–Id
Received	MES		Id
Passive	E	–Ign	
Active	E	Ign	

The reduced matrix has 16 elements of which 10 are non-trivial. The original matrix had 36 elements of which 20 were non-trivial.

By means of the reduced matrix, it is easy to check that the six place flows from Sect. 4.1 are really valid. To do this we show the place flows together with the incidence matrix. It turns out to be convenient to represent the place flows as columns – although they are really row vectors. It also turns out to be convenient to position the place flows to the left of the incidence matrix. To improve readability we list the range of each place flow at the top of the corresponding column.

Place Flows for Data Base System		PF_1	PF_2	PF_3	PF_4	PF_5	PF_6	T_1	T_3
		DBM	MES	E	DBM	MES	E	DBM	MES
Inactive	DBM	Id						–Id	–Rec
Waiting	DBM	Id				–Mes	–Ign	Id	
Performing	DBM	Id			Id				Rec
Unused	MES		Id					–Mes	
Sent, Ack.	MES		Id			Id		Mes	–Id
Received	MES		Id		–Rec	Id			Id
Passive	E			Id				–Ign	
Active	E			Id			Id	Ign	

To verify a place flow PF_i, we must show that the matrix product $PF_i * T_j$ is the zero function for all columns T_j in the matrix. To calculate the product of the two vectors imagine that the columns of PF_i and T_j are put next to each other (with PF_i to the left of T_j). Then combine each horizontal pair of elements by putting an $\circ$ operator between them and add the outcome as illustrated by the following example in which we check that $PF_5 * T_1 = 0$:

PF_5		T_1		$PF_5 * T_1$
MES		DBM		DBM→MES
		–Id		
–Mes		Id		–Mes ∘ Id
	*	–Mes	=	
Id		Mes		Id ∘ Mes
Id				
		–Ign		
		Ign		
				ZERO

As another and probably more convincing example, let us consider the telephone system from Sect. 3.2 of. Vol. 1. The incidence matrix has 13 rows and 14 columns, i.e., 182 elements of which 53 were non-trivial. However, it can be reduced to the following matrix with only 40 elements of which 13 are non-trivial. S denotes the function which maps a pair (s,r) into the sender s, while R denotes the function which maps (s,r) into the recipient r. In Sect. 4.5 we show how this reduction can be performed.

Reduced Matrix for Telephone System		T_1*	T_2*	T_3*	T_4*
		U	UxU	UxU	UxU
Inactive	U	–Id			
Continuous, Short, Disc.	U	Id	S	S+R	S+R
NoTone	U		–S		
Long	U			–S	
Ringing	U			–R	
Connected, Replaced	U				–S–R
Engaged	U	Id			
Request	UxU		–Id		
Call	UxU			–Id	
Connection	UxU				–Id

The transformations described in this section can be performed by a computer. The individual transformation rules are quite easy to implement and so is the translation between the graphical representation of a CP-net and the matrix representation. The only major problem is to find an efficient strategy for applying the transformation rules to sensible columns and rows in a sensible order. A discussion of this is outside the scope of this book.

The transformation rules are designed to work with incidence matrices for CP-nets. They exploit the fact that the incidence matrices are usually sparse matrices in which the same matrix elements occur many times. Moreover, there is often a high degree of linear dependency between the individual columns. The reduction rules are not complete, and it is easy to construct examples for which they provide little or no reduction. However, as indicated above, the transformations often yield a significant reduction of the incidence matrix. The reduction makes it easier to find place flows, because we get a simplified incidence matrix and hence a simplified matrix equation. The transformation rules can be used together with other methods to find place flows (see Sect. 4.3). This will often reduce the computational complexity of these methods considerably.

Computer tools for place invariant analysis

A CPN tool for place invariant analysis is being developed at Aarhus University, Denmark. The tool is closely integrated with the other CPN tools, i.e., the CPN editor and the CPN simulator described in Chap. 6 of Vol. 1 and the occurrence graph tool described in Chaps. 1–3.

The place invariant tool is able to find and check place invariants for hierarchical CP-nets. This is done by using the interactive approach described above – including the use of rules 1–7 to simplify the matrix and the use of lambda reductions to verify the flow property. The user specifies the place weights directly on the CPN diagram produced by the CPN simulator, i.e., by using the graphical representation of the CP-net.

The invariant tool is often able to deduce missing place weights from the other place weights. To illustrate this, let us again consider the data base system and let us assume that the modeller is looking for a place flow which relates the markings of *Inactive*, *Waiting*, and *Performing*. If the modeller specifies that all other places have weight zero, we may remove the corresponding rows from the incidence matrix – or any reduced matrix obtained from it. This can be done, because places with weight zero cannot contribute to the sums in the flow property and the invariant property. By removing the rows for *Unused*, *Sent*, *Received*, *Acknowledged*, *Passive*, and *Active* from the reduced matrix obtained above, the invariant tool gets the following matrix:

Reduced Matrix for Data Base System		T_1 DBM	T_3 MES
Inactive	DBM	–Id	–Rec
Waiting	DBM	Id	
Performing	DBM		Rec

By means of rule 5 (used on T_1 and T_3), the tool then obtains a "matrix" with one row and no columns. This means we are allowed to use any weight for the row. From the weight-factors of the row the tool deduces that:

$$W_{Inac} = W_{Wait} = W_{Perf}.$$

This means the invariant tool can tell the modeller that these three place weights must be identical – or the tool can calculate two of the place weights when the modeller specifies the third. If the modeller specifies that $W_{Inac} = Id$, the invariant tool calculates that $W_{Wait} = W_{Perf} = Id$. This means the place invariant PI_{DBM} has been found.

As another example, let us assume that the modeller is looking for a place invariant that tells him something about the relationship between the data base manager that is waiting and the messages that are in use, i.e., the messages on *Sent*, *Received*, and *Acknowledged*. The modeller may then specify that the remaining five places should have weight zero. By removing the corresponding rows from the reduced matrix obtained in Sect. 4.3, we get the following matrix:

Reduced Matrix for Data Base System		T_1 DBM	T_3 MES
Waiting	DBM	Id	
Sent, Ack.	MES	Mes	–Id
Received	MES		Id

By means of rule 7 (used on T_1 with $F = -Mes$) and rule 5 (used on T_3), we get a "matrix" with one row and no columns. This means we are allowed to use any weight for the row. From the weight-factors of the row the invariant tool deduces that:

$$W_{Sent} = W_{Rec} = W_{Ack}$$

$$W_{Wait} = -Mes \circ W_{Sent}.$$

By specifying a weight for one of the places the user obtains the place invariant PI_{W_A} (or a multiple of it).

As a third and final example, let us again consider the resource allocation system. We then have the following incidence matrix. The set P_Q denotes those elements of P for which the first component is q (if we delete the guard of T1 we can replace P_Q by P). Inc_Q is the function that maps any q-token (q,i) into (q,i+1) and maps any p-token into Ø. Inc_P is defined analogously.

Resource Allocation		T1	T2	T3	T4	T5
		P_Q	P	P	P	P
A	P	–Id				Inc_Q
B	P	Id	–Id			Inc_P
C	P		Id	–Id		
D	P			Id	–Id	
E	P				Id	–Id
R	E	$-Q_c$		Q_c		
S	E	$-Q_c$	$-2*P_c-Q_c$			$2*P_c+2*Q_c$
T	E			$-P_c$	$-P_c-Q_c$	$2*P_c+Q_c$

Let us now assume that the modeller is looking for a place invariant describing how r-resources are being used, while he does not care (for the moment) about the use of s-resources and t-resources. It is then sensible to specify that the place weight of R should be Id, while S and T have weight zero. Moreover, the modeller chooses some weight for A. It does not matter too much what weight he chooses for A, and hence we can just as well assume that he chooses a simple weight, such as the identity function or the zero function. We shall choose the last one (since it immediately leads to a simple place flow). With these assumptions, we can discard the rows for A, S, and T, to obtain the following matrix:

Resource Allocation		T1	T2	T3	T4	T5
		P_Q	P	P	P	P
B	P	Id	–Id			Inc_P
C	P		Id	–Id		
D	P			Id	–Id	
E	P				Id	–Id
R	E	$-Q_c$		Q_c		

which can be reduced to a "matrix" with one row and no columns. This means we are allowed to use any weight for the row. From the weight-factors of the row the invariant tool deduces that:

$$W_B = W_C = Q_c$$

$$W_D = W_E = Ø.$$

This means the place invariant PI_R has been constructed. The other three place invariants from Sect. 4.1 can be found in a similar way.

The place invariant tool is highly interactive. Typically the user defines a small number of non-zero weights (for places he is interested in) and a large number of zero weights (for places that are known to be without interest for the invariant he is constructing). Then the place invariant tool calculates those weights that can be uniquely determined from the weights proposed by the user. In this process the tool may also detect that some weights are inconsistent – because they violate the flow property. The situation can be compared to a spreadsheet model. Whenever the user changes a number in the spreadsheet, all depending equations are recalculated, and the new results are shown. Analogously, whenever the user changes one or more weights in the CP-net, all dependencies are recalculated, and the new derived weights are shown.

To calculate new weights and detect inconsistencies, the place invariant tool uses the reduction rules described above – but it reports the weights and the inconsistencies by showing them on the original CPN diagram. This can be done because Theorem 4.10 tells us how weights for a reduced matrix can be translated into weights for the original matrix. The user inspects the calculated place weights and the inconsistent transitions. Then he may add new place weights, modify existing place weights, or change the behaviour of transitions (e.g., by modifying arc expressions and guards). The process continues, with a number of iterations, and in the end a place invariant will be constructed (with some place weights specified by the user and the remaining ones derived by the place invariant tool).

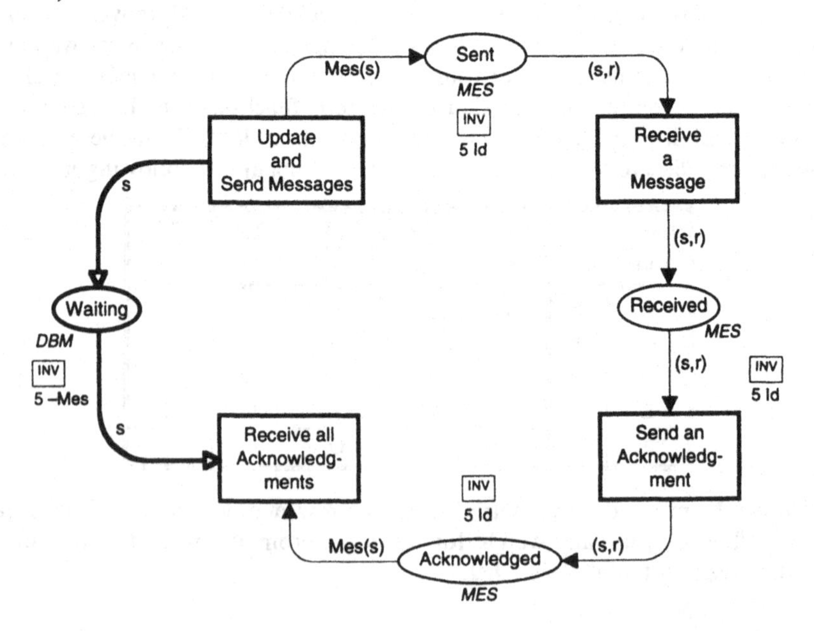

Fig. 4.2. Highlighted place invariant for the data base system

The method described above may seem primitive and cumbersome – but it is not. As illustrated by the examples above, it is often possible for the user to obtain useful place invariants by defining a few place weights. Over the last ten years we have taught the above method to students, who have successfully used it to find place invariants for small nets without having any computer support at all. The students have specified a set of proposed place weights, checked their consistency, and calculated derived weights. In practice, it always turned out that it was the last two things that were difficult, because they were time consuming and error-prone. With the place invariant tool this problem has disappeared because all the calculations are automatic. The check of a fully specified place invariant is a special case of the method described above. The user simply specifies *all* the place weights and the tool checks their consistency.

The invariant tool has several other features. It allows the modeller to construct invariants for hierarchical CP-nets by combining invariants for the individual pages. The tool also allows place invariants to be constructed as linear dependants of other place invariants. Finally, the tool allows the user to highlight an invariant by hiding all those places, transitions, and arcs which are of no interest due to zero weights. To highlight a place invariant the tool shows the subnet induced by those places that have a non-zero weight (see Def. 4.4 (ii) of Vol. 1). For the data base system the place invariant PI_{WA} is highlighted as shown in Fig. 4.2.

In the final version, the place invariant tool will also be able to assist the user when he uses invariants to prove dynamic properties of the modelled system. For example, the tool will use place invariants to derive upper and lower bounds for the marking of certain places, when the marking of other places are known. This will make the kind of arguments we made in Sect. 4.1 – to prove the dynamic properties of the data base system and the resource allocation system – much easier and less prone to error.

4.5 Transition Invariants

In this section we introduce transition invariants and transition flows. They are the duals of place invariants and place flows.

The intuition behind a **transition invariant** is simple. It is a step $Y \in BE_{MS}$ which has no effect, i.e., a step such that, for all places $p \in P$, we have that the added tokens are equal to the removed tokens (see Def. 2.9 of Vol. 1):

$$\sum_{(t,b)\in Y} E(p,t)<b> = \sum_{(t,b)\in Y} E(t,p)<b>.$$

This means an occurrence of the step Y will always produce a marking which is identical to the marking which existed prior to the occurrence of Y. The same is the case for all occurrence sequences which are obtained by splitting Y into a sequence of substeps.

A transition invariant Y is **executable** iff it can be split into a sequence of steps $Y_1, Y_2, \ldots, Y_k$ such that $Y = Y_1 + Y_2 + \ldots + Y_k$, $k \in \mathbb{N}$, and $M\,[Y_1, Y_2, \ldots, Y_k\rangle\, M$ for some $M \in [M_0\rangle$. The executable transition invariants can be used to find cyclic occurrence sequences that start in a reachable marking. Not all transition invariants are executable, because it may be the case that two binding elements are mutually dependent in the sense that each of them produces a token which the other one needs.

In the data base system there are many executable transition invariants. One example is the step TI_1 determined by:

$$TI_1 = Y_{1a} + Y_{2b} + Y_{3b} + Y_{4a} + Y_{5a}$$

where the steps to the right of the equality sign are those we considered in Sect. 2.3 of Vol. 1. There we saw that:

$$M_0 = M_1 [Y_{1a}\rangle M_2 [Y_{2b}\rangle M_3 [Y_{3b}\rangle M_4 [Y_{4a}\rangle M_5 [Y_{5a}\rangle M_6 = M_0.$$

The transition invariant TI_1 corresponds to the actions that are executed due to an update initiated by data base manager d_1. There is a similar transition invariant TI_i for each of the other data base managers $d_i \in \{d_2, \ldots, d_n\}$. It can be proved that the set $TI = \{TI_i \mid i \in 1..n\}$ generates all the transition invariants of the data base system. This means a step Y is a transition invariant iff it can be written as a sum $Y_1 + Y_2 + \ldots + Y_k$ where $k \in \mathbb{N}$ and $Y_j \in TI$ for all $j \in 1..k$.

As a second example, let us consider the telephone system from Sect. 3.2 of Vol. 1. It has many different transition flows, because there are many different action sequences by which a person may use a phone. One possibility is to lift the receiver and then immediately put it down again. Another possibility is to lift the receiver, dial a number, and then put the receiver down. For all the possible sequences which a user can enact, the phone system eventually ends up being in the same state as it was before the person started the call. This means the action sequence corresponds to a transition invariant. Below we list three of the transition invariants. From the cycles in the net structure (and the modeller's knowledge of the expected system behaviour) it is easy to find many others:

$$1`(\text{LiftInac},\langle x=u_3\rangle) + 1`(\text{RepCont},\langle x=u_3\rangle),$$

$$1`(\text{LiftInac},\langle x=u_3\rangle) + 1`(\text{Dial},\langle x=u_3, y=u_7\rangle) + 1`(\text{RepNoTo},\langle x=u_3, y=u_7\rangle),$$

$$\begin{aligned} &1`(\text{LiftInac},\langle x=u_3\rangle) + 1`(\text{Dial},\langle x=u_3, y=u_7\rangle) \\ &\quad + 1`(\text{Free},\langle x=u_3, y=u_7\rangle) + 1`(\text{LiftRinging},\langle x=u_3, y=u_7\rangle) \\ &\quad + 1`(\text{RepConSen},\langle x=u_3, y=u_7\rangle) + 1`(\text{RepDisc},\langle y=u_7\rangle). \end{aligned}$$

Next let us consider transition flows. In Sects. 4.2 and 4.3 we saw that a place flow (with range $A \in \Sigma$) is a solution to the matrix equation:

$$W * I = 0$$

where the unknown W is a row vector such that $W_p \in [C(p)_{WS} \rightarrow A_{WS}]_L$ for all $p \in P$. Analogously, we shall define a **transition flow** (with domain $A \in \Sigma$) to be a solution to the matrix equation:

$$I * W = 0$$

where the unknown W is a column vector such that $W_t \in [A_{WS} \rightarrow B(t)_{WS}]_L$ for all $t \in T$. The matrix product $I * W$ is defined in a similar way to $W * I$, i.e., by using functional composition as the multiplication operator. The function W_t is said to be the **weight** of the transition t.

As for place flows, we say that a transition flow is non-negative iff all the transition weights are non-negative functions. The intuition behind a non-negative transition flow W is that it determines, for each $a \in A$ in the domain of W, a transition invariant $W(a) \in BE_{MS}$ defined by:

$$W(a)(t,b) = W_t(a)(b)$$

for all $(t,b) \in BE$. This can be seen from the following sequence of calculations, which is valid for all $a \in A$ and all $p \in P$ (explanation follows below):

$$\begin{aligned}(I * W)(p)(a) &= \sum_{t \in T} (I(p,t))(W_t(a)) \\ &= \sum_{t \in T} \sum_{b \in W_t(a)} (I(p,t))(b) \\ &= \sum_{(t,b) \in W(a)} (I(p,t))(b) \\ &= \sum_{(t,b) \in W(a)} E(t,p)<b> - \sum_{(t,b) \in W(a)} E(p,t)<b>.\end{aligned}$$

The first equality follows from the definition of the matrix product $I * W$. The second equality follows from the definition of linear functions between weighted-sets. The third equality follows from the definition of W(a), while the last follows from the definition of I(p,t).

From the calculations above, it follows that a set of non-negative transition weights W is a transition flow iff W(a) is a transition invariant for all $a \in A$. This means we can use transition flows to find transition invariants – in a similar way as we use place flows to find place invariants.

For the data base system we can determine the n transition invariants in the generating set $\{TI_1, TI_2, \ldots, TI_n\}$ from the transition flow which has domain DBM and weights:

$$W_{SM} = W_{RA} = Id$$

$$W_{RM} = W_{SA} = Mes.$$

To see that W really is a transition flow, we check that $I * W = 0$, i.e., that:

SM	RA	RM	SA
DBM	DBM	MES	MES
–Id	Id	–Rec	Rec
Id	–Id		
		Rec	–Rec
–Mes	Mes		
Mes		–Id	
		Id	–Id
	–Mes		Id
–Ign	Ign		
Ign	–Ign		

∗ (Id, Id, Mes, Mes) = 0

As with place flows, it is convenient to represent the transition flows directly on the CP-net. This is illustrated by Fig. 4.3 in which we show seven different transition flows for the telephone system from Sect. 3.2 of Vol. 1. The transition flows determine the transition invariants discussed above. All the flows are non-negative and hence we can use them to determine transition invariants.

Above we have given the intuition behind transition flows and transition invariants. Now we give the formal definitions – first for a non-hierarchical CP-net, then for a hierarchical net. Note that the empty multi-set over BE is always a transition invariant.

Definition 4.11: For a non-hierarchical CP-net, a **set of transition weights** with domain $A \in \Sigma$ is a set of functions $W = \{W_t\}_{t\in T}$ such that $W_t \in [A_{WS} \to B(t)_{WS}]_L$ for all $t \in T$.

(i) W is a **transition flow** iff:

$$\forall a\in A\ \forall p\in P: \sum_{(t,b)\in W(a)} E(p,t)\langle b\rangle = \sum_{(t,b)\in W(a)} E(t,p)\langle b\rangle.$$

(ii) A finite multi-set $Y \in BE_{MS}$ is a **transition invariant** iff:

$$\forall p\in P: \sum_{(t,b)\in Y} E(p,t)\langle b\rangle = \sum_{(t,b)\in Y} E(t,p)\langle b\rangle.$$

Theorem 4.12: For a set of non-negative transition weights W, with domain A, we have:

W is a transition flow $\Leftrightarrow$ $\forall a \in A$: W(a) is a transition invariant.

Proof: Straightforward consequence of Def. 4.11. □

Definition 4.13: For a hierarchical CP-net, a **set of transition weights** with domain $A \in \Sigma$ is a set of functions $W = \{W_{t'}\}_{t' \in TI}$ such that $W_{t'} \in [A_{WS} \rightarrow B(t')_{WS}]_L$ for all $t' \in TI$.

(i) W is a **transition flow** iff:

$$\forall a \in A\ \forall p'' \in PIG: \sum_{\substack{(t',b) \in W(a) \\ p' \in p''}} E(p',t')<b> = \sum_{\substack{(t',b) \in W(a) \\ p' \in p''}} E(t',p')<b>.$$

(i) A finite multi-set $Y \in BE_{MS}$ is a **transition invariant** iff:

$$\forall p'' \in PIG: \sum_{\substack{(t',b) \in Y \\ p' \in p''}} E(p',t')<b> = \sum_{\substack{(t',b) \in Y \\ p' \in p''}} E(t',p')<b>.$$

Note that a transition invariant (in contrast to a place invariant) is a static property that can be checked without considering the set of all reachable markings. Theorem 4.12 remains valid when we go from non-hierarchical CP-nets to hierarchical nets (i.e., replace Def. 4.11 by Def. 4.13).

Transition flows can be constructed/calculated in a similar way to place flows, i.e., by modifications of the automatic and interactive techniques described in Sects. 4.3 and 4.4. The tool support is very similar to that of place invariants and hence we shall not describe it any further. The transformation rules which preserve transition flows deal with rows instead of columns, but otherwise they are analogous to the rules in Sect. 4.4.

Transition invariants can be used to prove different kinds of system properties. As an example, we can use the executable transition invariants to investigate fairness properties, because they give us information about cyclic occurrence sequences.

One of the reasons for introducing transition flows is the fact that they can be used to calculate place flows, via the reduction rules in Sect. 4.4. To see how this can be done, let us again consider the data base system and the transition flow:

$$W_{SM} = W_{RA} = Id$$

$$W_{RM} = W_{SA} = Mes.$$

From $I * W = 0$ we conclude that the following linear combination (of columns in the incidence matrix):

$$SM + RA + RM \circ Mes + SA \circ Mes$$

must be a column in which all elements are zero functions. This implies that *RA* is linearly dependent on the other three columns. Hence we can use rule 3 of Sect. 4.4 to remove *RA* without changing the set of *place* flows. Instead, we could have removed *SM*, but we could not have removed *RM* or *SA* (because *Mes* is not pseudo-surjective).

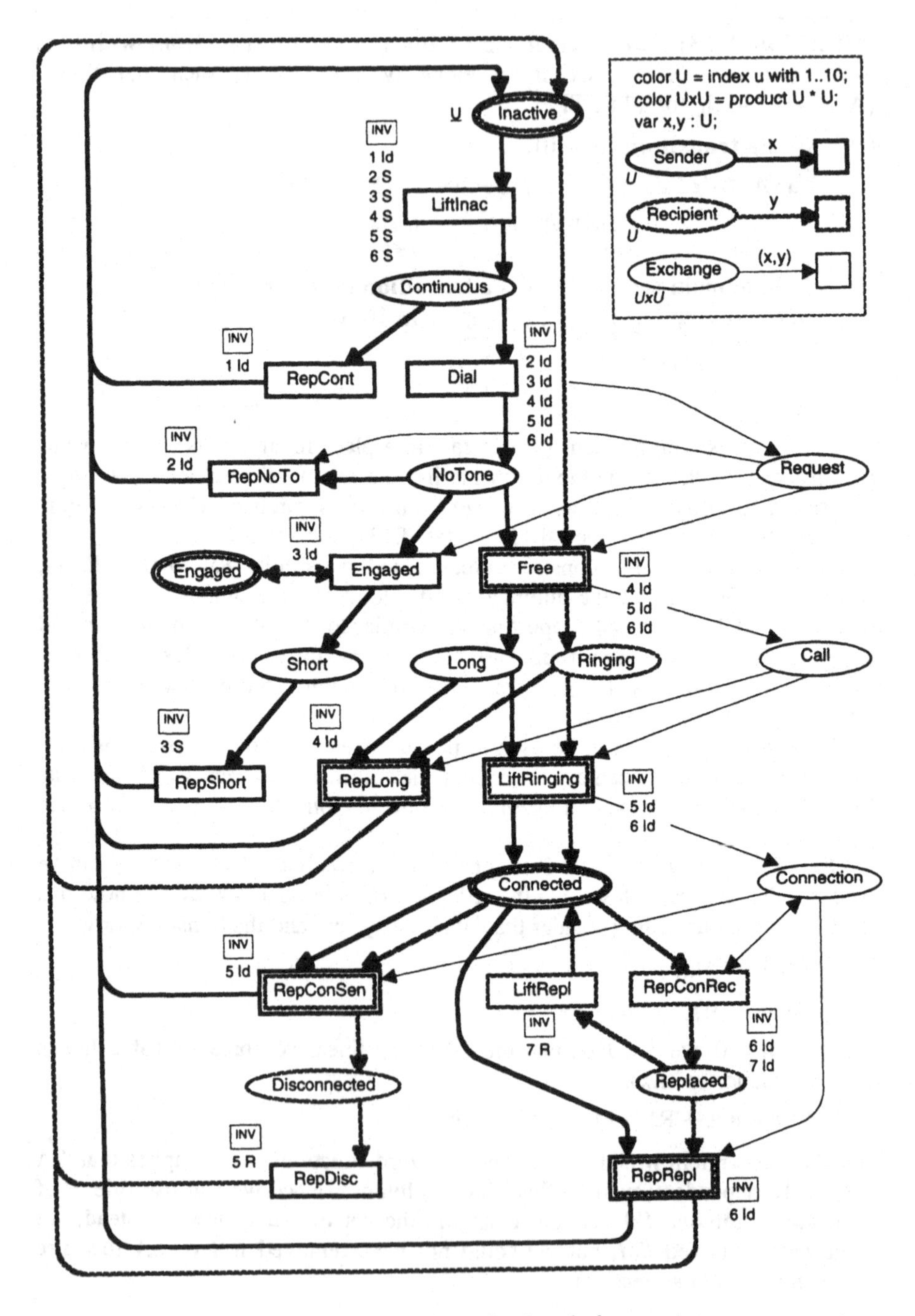

Fig. 4.3. Transition flows for the telephone system

As a second example, let us consider the telephone system and the transition flows from Fig. 4.3. The first transition flow allows us to remove the column for *RepCont*. This can be done because the flow tells us that the column for *RepCont* is linearly dependent on the column for *LiftInac*. We could also have removed the column for *LiftInac*, but we choose *RepCont* because the latter is not used in any of the other transition flows. By means of the other six transition flows we remove the columns for *RepNoTone, Engaged, RepLong, RepConSen, RepRepl*, and *RepConRec*, respectively. Then we have the following matrix:

Reduced Matrix for Telephone System		Lift Inac	Dial	Free	Rep Short	Lift Ring	Rep Disc	Lift Repl
		T_1	T_2	T_3	T_4	T_5	T_6	T_7
		U	UxU	UxU	U	UxU	U	U
Inactive	U	–Id		–R	Id		Id	
Continuous	U	Id	–S					
NoTone	U		S	–S				
Short	U				–Id			
Long	U			S		–S		
Ringing	U			R		–R		
Connected	U					S+R		Id
Disconnected	U						–Id	
Replaced	U							–Id
Engaged	U	Id		R	–Id		–Id	
Request	UxU		Id	–Id				
Call	UxU			Id		–Id		
Connection	UxU					Id		

By adding T_1 to T_4 and by adding T_1 to T_6, we get two columns for which we can use rule 5. Moreover, we can use rule 5 on T_7. We then obtain the following matrix:

Reduced Matrix for Telephone System		T_1	T_2	T_3	T_5
		U	UxU	UxU	UxU
Inactive	U	–Id		–R	
Continuous, Short, Disc.	U	Id	–S		
NoTone	U		S	–S	
Long	U			S	–S
Ringing	U			R	–R
Connected, Replaced	U				S+R
Engaged	U	Id		R	
Request	UxU		Id	–Id	
Call	UxU			Id	–Id
Connection	UxU				Id

which we can beautify – by means of rules 1 and 2 of Sect. 4.4 – to obtain the reduced matrix presented in Sect. 4.4:

Reduced Matrix for Telephone System		T_1*	T_2*	T_3*	T_4*
		U	UxU	UxU	UxU
Inactive	U	–Id			
Continuous, Short, Disc.	U	Id	S	S+R	S+R
NoTone	U		–S		
Long	U			–S	
Ringing	U			–R	
Connected, Replaced	U				–S–R
Engaged	U	Id			
Request	UxU		–Id		
Call	UxU			–Id	
Connection	UxU				–Id

Above we have shown how to reduce the incidence matrix of the telephone system. It should be noted that the reduction can be done automatically as soon as the seven transition flows are known. The transition flows can be checked automatically, and they are straightforward to find, because they directly follow from the modeller's understanding of some of the basic system properties.

4.6 Uniform CP-nets

In this section we consider a particular kind of CP-nets called uniform CP-nets. For each uniform CP-net we can construct a PT-net that can be used to determine place and transition flows of the uniform net. The constructed PT-net has the same net structure as the CP-net and hence it is usually much smaller than the equivalent PT-net (see Sect. 2.4 of Vol. 1).

A **uniform** CP-net is a net in which all transitions are uniform (see Def. 4.5 of Vol. 1). This means each transition always move the same number of tokens along a given arc, independently of the occurring binding. When a CP-net is uniform and the initial marking $M_0 \in TE_{MS}$ is finite, it makes sense to define the **underlying PT-net**. This is done by replacing each arc expression E(a) by $|E(a)|$ and each initialization expression I(p) by $|I(p)|$. Simultaneously, we discard all colours sets and all guards. For the data base system we get the PT-net shown in Fig. 4.4 (omitted arc expressions are equal to 1).

Definition 4.14: Let a uniform non-hierarchical CP-net CPN = (Σ, P, T, A, N, C, G, E, I) with finite initial marking be given. Then we define the **underlying PT-net** to be PTN* = (P*, T*, A*, E*, I*) where:

(i) $P^* = P$.

(ii) $T^* = T$.

(iii) $A^* = \{(p,t) \in P^* \times T^* \mid \sum_{a \in A(p,t)} |E(a)| \neq 0\} \cup \{(t,p) \in T^* \times P^* \mid \sum_{a \in A(t,p)} |E(a)| \neq 0\}$.

(iv) $\forall (x_1,x_2) \in A^*: E^*(x_1,x_2) = \sum_{a \in A(x_1,x_2)} |E(a)|$.

(v) $\forall p \in P^*: I^*(p) = |I(p)|$.

A place weight $W_p \in [C(p)_{WS} \rightarrow A_{WS}]_L$ is said to be **uniform** with multiplicity $z \in \mathbb{Z}$ iff $|W_p(c)| = z$ for all $c \in C(p)$. We use $|W_p|$ to denote the multiplicity of W_p. For a set of uniform place weights $W = \{W_p\}_{p \in P}$, we use $|W|$ to denote the integer weights $\{|W_p|\}_{p \in P}$. Analogous definitions are made for transition weights.

Next we define place flows and place invariants for PT-nets. We require all weights to be integers and * denotes standard integer multiplication:

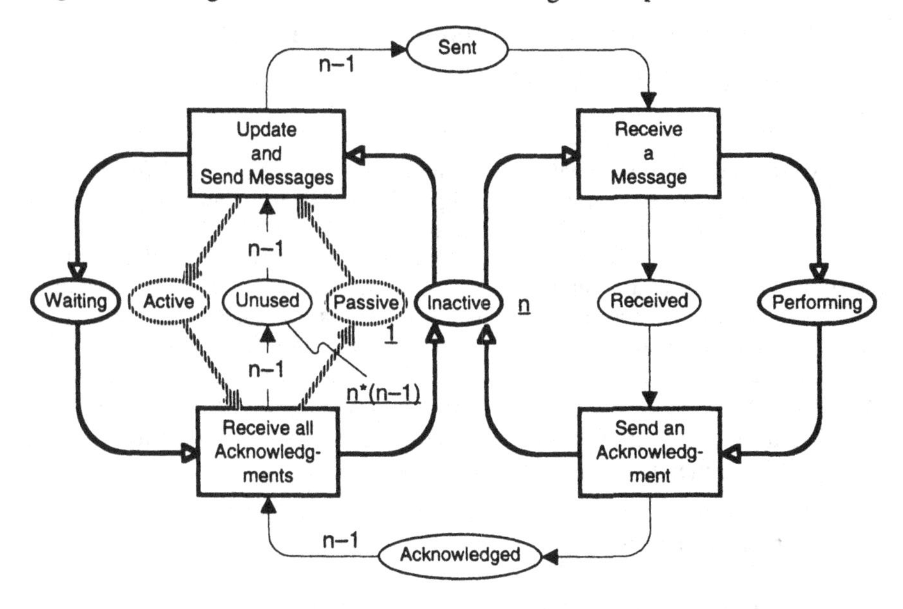

Fig. 4.4. Underlying PT-net for the data base system

Definition 4.15: For a PT-net, a **set of place weights** is a set $W = \{W_p\}_{p \in P}$ such that $W_p \in \mathbb{Z}$ for all $p \in P$.

(i) W is a **place flow** iff:

$$\forall t \in T: \sum_{p \in P} W_p * E(p,t) = \sum_{p \in P} W_p * E(t,p).$$

(ii) W determines a **place invariant** iff:

$$\forall M \in [M_0\rangle: \sum_{p \in P} W_p * M(p) = \sum_{p \in P} W_p * M_0(p).$$

It is easy to show that the analogue of Theorem 4.7 is also valid for PT-nets – replace "no dead binding elements" by "no dead transitions".

Definition 4.16: For a PT-net, a **set of transition weights** is a set $W = \{W_t\}_{t \in T}$ such that $W_t \in \mathbb{Z}$ for all $t \in T$.

(i) W is a **transition flow** iff:

$$\forall p \in P: \sum_{t \in T} W_t * E(p,t) = \sum_{t \in T} W_t * E(t,p).$$

(ii) A finite multi-set $Y \in T_{MS}$ is a **transition invariant** iff:

$$\forall p \in P: \sum_{t \in Y} E(p,t) = \sum_{t \in Y} E(t,p).$$

It is easy to show that the analogue of Theorem 4.12 is also valid for PT-nets. Each set of non-negative transition weights $W = \{W_t\}_{t \in T}$ determines a *single* transition invariant in which the coefficient of t is W_t for all $t \in T$.

Now let us consider the matrix representation of a PT-net, i.e., the incidence matrix and the row and column vectors for markings, steps and sets of weights. The matrix representation is totally analogous to that of a CP-net – but now we have much simpler matrix elements. All of them are integers.

Theorem 4.17: Let CPN be a uniform non-hierarchical CP-net with finite initial marking and let PTN be the underlying PT-net. For a uniform set of weights W, we then have the following properties:

(i) W place flow for CPN $\Rightarrow$ $|W|$ place flow for PTN.
(ii) W transition flow for CPN $\Rightarrow$ $|W|$ transition flow for PTN.

Proof: To prove (i), we assume that W is a place flow for CPN, i.e., that:

$$(W * I)(t) = 0$$

for all $t \in T$. From the definition of the matrix product in Sect. 4.3 and from the definition of uniform weights and arc expressions we get the following sequence of calculations:

$$|(W * I)(t)| = |\sum_{p \in P} W_p \circ I(p,t)| = \sum_{p \in P} |W_p| * |I(p,t)| = (|W| * |I|)(t)$$

where $|I|$ is the incidence matrix of PTN. Hence we conclude that:

$$(|W| * |I|)(t) = 0$$

for all $t \in T$, i.e., that $|W|$ is a place flow for PTN. The proof of (ii) is analogous.
□

Theorem 4.17 is important because it allows us to check whether a set of place invariants for a uniform CP-net is "complete" or not. For each known CP-net flow W that is uniform, we construct a PT-net flow $|W|$. Then we check whether the constructed PT-net flows form a basis for the solutions to $W * |I| = 0$. If some of the constructed PT-net flows are linearly dependent on each other, we can usually discard one or more of the corresponding CP-net flows – without losing information. If there are too few PT-net flows to form a basis, we can find some more, and use these to construct CP-net flows. To construct a CP-net flow W from a PT-net flow U, we have to choose the weights of W such that each function W_p is uniform with multiplicity U_p. One possible choice is to use $W_p = U_p * \mathrm{Ign}$. This will always work, but usually we can choose better weights yielding place invariants with more information.

To illustrate the technique, let us consider the data base system. The underlying PT-net has the following incidence matrix:

Data Base System	SM	RA	RM	SA
Inactive	−1	1	−1	1
Waiting	1	−1		
Performing			1	−1
Unused	−(n−1)	n−1		
Sent	n−1		−1	
Received			1	−1
Acknowledged		−(n−1)		1
Passive	−1	1		
Active	1	−1		

It is easy to verify that the incidence matrix has rank 3. The rank is the number of linearly independent columns. It is known to be identical to the number of linearly independent rows. From linear algebra, we then know that the matrix equation $W * |I| = 0$ has:

$$|P^*| - \mathrm{rank}(|I|) = 9 - 3 = 6$$

linearly independent solutions. Hence we conclude that we can expect to be able to find six "independent" place flows for the data base system. This is consistent with the results in Sect. 4.1. Analogously, we can expect to find:

$$|T^*| - \mathrm{rank}(|I|) = 4 - 3 = 1$$

"independent" transition flows. This is consistent with the results of Sect. 4.5.

As a second example, let us consider the telephone system. Instead of constructing the incidence matrix for the underlying PT-net, we can start from the reduced matrix shown below (to the right).

Place Flows for Telephones		PF_1	PF_2	PF_3	PF_4	PF_5	PF_6	T_1*	T_2*	T_3*	T_4*
		U	U	U	U	U	U	U	UxU	UxU	UxU
Inactive	U	Id	Id					–Id			
Cont., Sh., Disc.	U	Id						Id	S	S+R	S+R
NoTone	U	Id		Id					–S		
Long	U	Id			Id					–S	
Ringing	U	Id				Id				–R	
Connect., Repl.	U	Id					Id				–S–R
Engaged	U		Id					Id			
Request	UxU			–S					–Id		
Call	UxU				–S	–R				–Id	
Connection	UxU						–S–R				–Id

The underlying matrix has rank 4. Hence we can expect to find 10–4=6 "independent" place flows. One such possible set is determined by the six place flows shown above (in the middle).

Let us finish the section by observing that the above method can also be used for hierarchical CP-nets. Then we either consider the non-hierarchical equivalent, or we consider each page at a time (see the last paragraph of Sect. 4.3). The method can also be used for a "nearly uniform" CP-net. This is done by splitting each transition t that has a non-uniform arc a. We then get a transition for each member of the set $\{ |E(a)(b)| \mid b \in B(t)\}$. For this to be possible the set must be finite and, in practice, it must also be small.

Bibliographical Remarks

As explained in the bibliographical remarks to Chap. 1 of Vol. 1, an improved handling of the invariant method was the main motivation behind the development of the first kind of CP-nets. The basic ideas behind the matrix representation were developed in [22]. The flow-preserving reduction rules are described in [23] together with a proof of their soundness. The paper also defines uniform CP-nets, introduces the underlying PT-net, and establishes the connection between the place flows of these nets.

Different methods for automatic calculation of place flows can be found in [13], [14], [31], [39], and [41]. Modular calculation of place flows are discussed in [9] and [32].

An introduction to lambda expressions and lambda reductions can be found in [17] and [38]. A description of the place invariant tool from Sect. 4.4 can be found in [10].

Exercises

Exercise 4.1.
Consider the philosopher system from Exercise 1.6 of Vol. 1 (or Sect. 1.6).

(a) Construct some place flows for the philosopher system. Verify that the place flows are correct.

(b) Use the place flows to investigate the boundedness properties of the philosopher system.

(c) Use the place flows to investigate the home and liveness properties of the philosopher system.

(d) Use the place flows to prove that two neighbouring philosophers cannot eat at the same time.

(e) Use the results in Sect. 4.6 to discuss whether you should expect to be able to find additional place flows for the philosopher system.

Exercise 4.2.
Consider the telephone system from Sect. 3.2 of Vol. 1. For this system we have determined seven transition flows (in Sect. 4.5) and six place flows (in Sect. 4.6).

(a) Use the place flows to verify the upper bounds postulated at the end of Sect. 4.2 of Vol. 1.

(b) Use the place flows to verify that the initial marking is a home marking.

(c) Use the place flows to investigate the liveness properties of the telephone system.

(d) Use the transition flows to investigate the fairness properties of the telephone system.

Exercise 4.3.
Consider the gas pump system from Exercise 1.7 of Vol. 1.

(a) Construct some place flows for the gas pump system. Verify that the place flows are correct.

(b) Use the place flows to investigate the boundedness properties of the gas pump system.

(c) Use the place flows to investigate the home and liveness properties of the gas pump system.

(d) Construct some transition flows for the gas pump system. Verify that the transition flows are correct.

(e) Use the transition flows to investigate the fairness properties of the gas pump system.

(f) Use the results in Sect. 4.6 to discuss whether you should expect to be able to find additional flows for the gas pump system.

Exercise 4.4.
Consider the master/slave system from Exercise 1.9 of Vol. 1.

(a) Construct some place flows for the master/slave system. Verify that the place flows are correct.

(b) Use the place flows to investigate the boundedness properties of the master/slave system.

(c) Use the place flows to investigate the home and liveness properties of the master/slave system.

(d) Construct some transition flows for the master/slave system. Verify that the transition flows are correct.

(e) Use the transition flows to investigate the fairness properties of the master/slave system.

(f) Use the results in Sect. 4.6 to discuss whether you should expect to be able to find additional flows for the master/slave system.

Exercise 4.5.
Consider the process control system from Exercise 4.6 of Vol. 1.

(a) Construct some place flows for the process control system. Verify that the place flows are correct.

(b) Use the place flows to investigate the boundedness properties of the process control system.

(c) Use the place flows to investigate the home and liveness properties of the process control system.

(d) Construct some transition flows for the process control system. Verify that the transition flows are correct.

(e) Use the transition flows to investigate the fairness properties of the process control system.

(f) Use the results in Sect. 4.6 to discuss whether you should expect to be able to find additional flows for the process control system.

Exercise 4.6.
Consider the flow-preserving transformation rules defined in Sect. 4.4.

(a) Fill in the missing details of the proof for Theorem 4.10.

(b) Construct a set of transformation rules which preserve transition flows.

Exercise 4.7.
Consider the data base system and the transition invariants $\{TI_1, TI_2, \ldots, TI_n\}$ determined in Sect. 4.5.

(a) Prove that the elements of $\{TI_1, TI_2, \ldots, TI_n\}$ generate all transition invariants of the data base system.

Chapter 5

Timed CP-nets

Most applications of CP-nets are used to investigate the logical correctness of a system. This means we consider the dynamic properties and the functionality of the system. However, CP-nets can also be used to investigate the performance of a system, e.g., the maximal time used for the execution of certain activities and the average waiting time for certain requests. To perform this kind of analysis, we extend the CPN model with a time concept, represented by a global clock (which may be continuous or discrete). We then specify how the different activities and states "consume" time. To do this, we allow each token to carry a time stamp – in addition to the token colour. The time stamp tells us when the token is ready to be used by a transition. When a token is created, the time stamp is specified by an expression. This means it is possible to specify all kinds of delays (constant, interval, or probability distribution). It also means that the delay may depend upon the binding of the transition which creates the token.

There are several other ways in which CP-nets can be extended with a time concept. A brief discussion can be found in the bibliographical remarks.

Section 5.1 contains an informal introduction to timed multi-sets and timed CP-nets. This is done by means of the resource allocation system. Section 5.2 contains the formal definition of timed multi-sets. They constitute a straightforward modification of the ordinary multi-sets defined in Chap. 2 of Vol. 1. Section 5.3 contains the formal definition of timed CP-nets and their behaviour. The definitions are straightforward modifications of the corresponding definitions for untimed CP-nets in Chap. 2 of Vol. 1. Section 5.4 discusses the relationships between timed and untimed CP-nets. It turns out that each timed CP-net determines an untimed CP-net, such that the set of occurrence sequences of the timed net is a subset of the set of occurrence sequences of the untimed net. Intuitively, this means the two nets have the same behaviour, except that the time constraints in the timed net may rule out some of the occurrence sequences which are legal in the untimed net. Section 5.5 contains a second example of timed CP-nets. There, we consider a simple protocol. Finally, Sect. 5.6 discusses how to analyse the behaviour of timed CP-nets – by means of simulation, occurrence graphs, and invariants.

5.1 Introduction to Timed CP-nets

To investigate the performance of systems, i.e., the speed at which they operate, it is convenient to extend CP-nets with a time concept. To do this, we introduce a **global clock**. The clock values represent the **model time**, and they may either be discrete (e.g., integers) or continuous (e.g., reals). In addition to the token colour, we allow each token to carry a **time value**, also called a **time stamp**. Intuitively, the time stamp describes the *earliest* model time at which the token can be used, i.e., removed by a binding element.

In a timed CP-net a binding element is said to be **colour enabled** when it satisfies the requirements of the usual enabling rule (Defs. 2.8 and 3.6 of Vol. 1). However, to be **enabled**, the binding element must also be **ready**. This means all the time stamps of the removed tokens must be less than or equal to the current model time (removed tokens are defined in Defs. 2.9 and 3.6 of Vol. 1).

To model that an activity/operation takes Δr time units, we let the corresponding transition t create time stamps for its output tokens that are Δr time units larger than the clock value at which t occurs. This implies that the tokens produced by t are unavailable for Δr time units. It can be argued that it would be more natural to delay the creation of the output tokens, so that they did not come into existence until Δr time units after the occurrence of t had begun. However, such an approach would mean that a timed CP-net would get "intermediate" markings which do not correspond to markings in the corresponding untimed CP-net, because there would be markings in which input tokens have been removed but output tokens not yet generated. Hence we would get a more complex relationship between the behaviour of timed and untimed nets.

The execution of a timed CP-net is time driven, and it works in a similar way to that of the event queues found in many programming languages for discrete event simulation. The system remains at a given model time as long as there are colour enabled binding elements that are ready for execution. When no more binding elements can be executed, at the current model time, the system advances the clock to the next model time at which binding elements can be executed. Each marking exists in a closed interval of model time (which may be a point, i.e., a single moment). The occurrence of a binding element is instantaneous.

To see how a timed simulation works, let us consider Fig. 5.1, which contains a timed CP-net for the resource allocation system from Sect. 1.2 of Vol. 1. For this system, we use a discrete clock, starting at 0. From the third and fourth lines of the declarations, we see that P-tokens are timed (i.e., carry time stamps), while the E-tokens are not. This means E-tokens are always ready to be used. The small rectangle below the declarations indicates that the current model time is 641 (in the CPN simulator this information is displayed in the status bar).

The @ signs in the current markings should be read "at". Each @ sign is followed by a list of time stamps. The marking of place A contains two tokens, one with colour (q,4) and time stamp 627, and one with colour (q,5) and time stamp 598. Analogously, place B has two tokens, one with colour (p,6) and time stamp 567, and one with colour (q,1) and time stamp 602. Place D has a single token with colour (p,4) and time stamp 641. Finally, place T has one token with colour e and no time stamp. In Fig. 5.1 all lists of time stamps have length 1, because all the tokens have different colours. However, the initial marking of place A is displayed as 3`(q,0)@[0,0,0], while the initial marking of B is 2`(p,0)@[0,0].

Now let us consider the steps which may occur in the marking of Fig. 5.1. The binding element b_1 = (T4,<x=p, i=4>) is colour enabled, because the two input places have the necessary tokens. The binding element is also ready, because all the time stamps of the removed tokens are smaller than or equal to the current model time (in this case there is only one such time stamp and it is equal to the current model time). Hence b_1 is enabled and it may occur. The occurrence of b_1 removes the token from D and the token from T. The occurrence of b_1 will also add a token to E. The colour of the new token is calculated in the usual way, i.e., as specified by the occurrence rule (Defs. 2.9 and 3.6 of Vol. 1). The time stamp of the new token is calculated as the current model time plus a time delay, which is specified in the corresponding output arc expression – after the @+ operator. In this case the delay is 12 time units, and hence we get 653 as the new time stamp. Intuitively, this means the p-token must stay at least 12 time

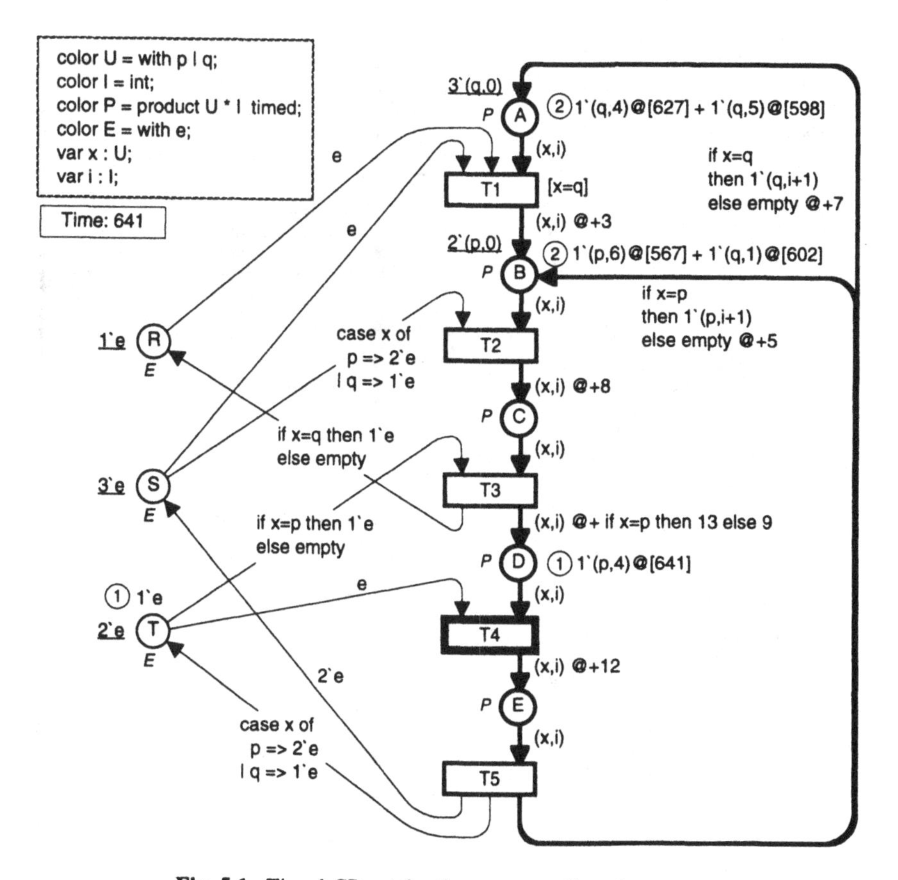

Fig. 5.1. Timed CP-net for the resource allocation system

units at place E. We can interpret this to mean that the state E has a minimal duration of 12 time units. We can also interpret it to mean that the activity T4 takes 12 time units.

When b_1 has occurred, we reach a marking in which b_2 = (T5,<x=p, i=4>) is the only colour enabled binding element. However, b_2 is not ready at model time 641 because it involves a token with a too-high time stamp. Hence we increase the model time until b_2 becomes ready, i.e., to 653. If there had been several colour enabled binding elements we would have increased the model time until one of them became ready. When b_2 has occurred, at model time 653, we get the marking shown in Fig. 5.2. Now we have three colour enabled binding elements b_3 = (T2,<x=p, i=5>), b_4 = (T2,<x=p, i=6>), and b_5 = (T2,<x=q, i=1>). The binding elements b_4 and b_5 are ready at 653, while b_3 uses a token with time stamp 658. Hence either b_4 or b_5 will occur. They are in conflict with each other, because they both need e-tokens from S. This means only one of them will be executed. However, had there been an additional e-token on S, b_4 and b_5 would have been concurrently enabled and both of them would have occurred at time 653 (either in the same step, or in two subsequent steps).

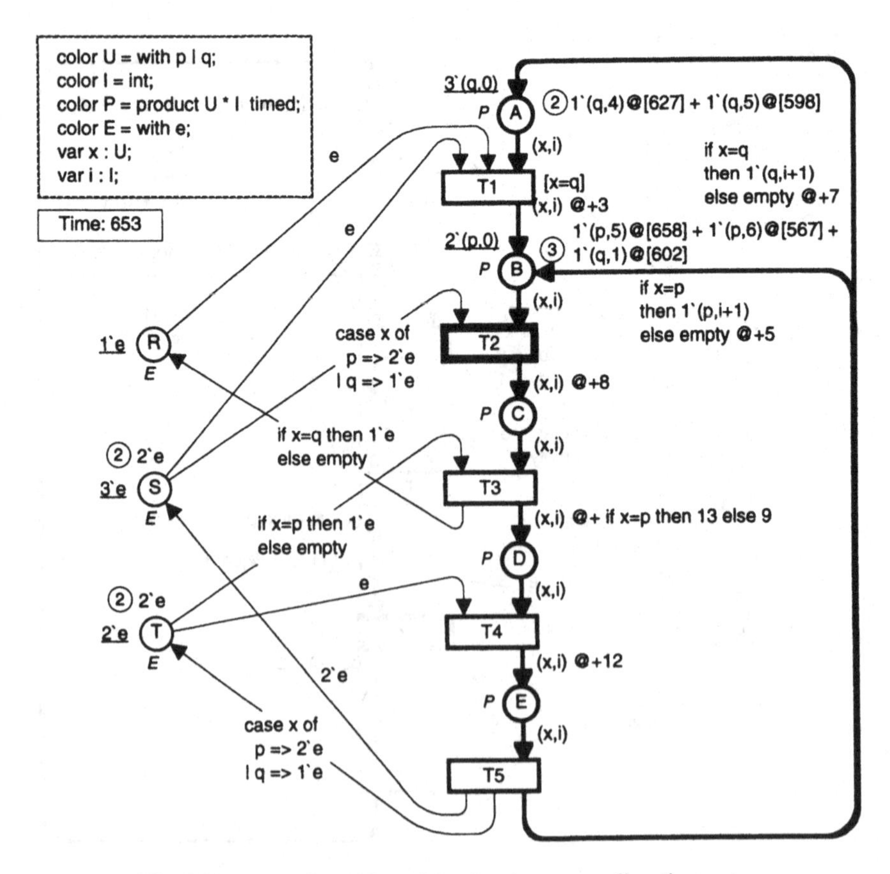

Fig. 5.2. A second marking of the timed resource allocation system

The time delays may depend upon the binding, i.e., upon the colours of the input and output tokens. This is illustrated by the output arc of T3, where the delay depends upon the value of the variable x. For p-processes the delay is 13, while for q-processes it is 9. The delays are specified by means of expressions, and this means, for example, they can use functions which implement complex statistical distributions.

For a timed CP-net we require that each step consists of binding elements which are both colour enabled and ready. Hence the possible occurrence sequences of a timed CP-net always form a subset of the possible occurrence sequences of the corresponding untimed CP-net. This means we have a well defined and easy-to-understand relationship between the behaviour of a timed CP-net and the behaviour of the corresponding untimed CP-net.

In the resource allocation system, we have only illustrated one of the simplest ways in which time stamps can be used. All removed tokens for a binding element (t,b) were required either to be without time stamps or to have time stamps which were less than or equal to the time value r^* at which (t,b) occurs. At a given place p, all added tokens for (t,b) either got no time stamps or got identical time stamps which were equal to r^* plus a delay Δr. In general, the situation can be considerably more complex. We allow each arc expression to evaluate to a timed multi-set, i.e., a multi-set of pairs (c,r) where c is a token colour while r is a time value. This means different time constraints may be defined for each individual token involved in the occurrence of (t,b).

When an input arc expression evaluates to a timed multi-set $(c_1,r_1), (c_2,r_2), \ldots, (c_n,r_n)$ (possibly with repetitions) we require the corresponding input place to contain (at least) a multi-set of tokens $(c_1,r_1^*), (c_2,r_2^*), \ldots, (c_n,r_n^*)$ such that $r_i^* \le r^*+r_i$ for all $i \in 1..n$. When an output arc expression evaluates to a timed multi-set $(c_1,r_1), (c_2,r_2), \ldots, (c_n,r_n)$ (possibly with repetitions), the corresponding output place get a multi-set of tokens $(c_1,r_1^*), (c_2,r_2^*), \ldots, (c_n,r_n^*)$ such that $r_i^* = r^*+r_i$ for all $i \in 1..n$.

A positive time value in an output arc expression implies that the corresponding token gets a time stamp that is larger than the current model time. Intuitively this means the activity/operation modelled by the transition takes time. A positive time value in an input arc expression implies that we allow the corresponding token to be used a specified amount of time ahead of its time stamp. This can be extremely useful. As an example, a token may model a timer, expiring at the time stamp. By using a positive time value in an input arc expression, it is possible to stop the timer, i.e., remove the token ahead of the time stamp.

Let us now consider how the notation of Fig. 5.1 fits into the more general framework described above. Each input arc expression of Fig. 5.1 evaluates to an ordinary, untimed, multi-set $c_1, c_2, \ldots, c_n$. For the places A–E, with a timed colour set, we consider this to be a shorthand for the timed multi-set $(c_1,0), (c_2,0), \ldots, (c_n,0)$. The output arc expressions use the dyadic operator @+. The operator takes a multi-set m and a time value r, and it returns the timed multi-set which consists of all the elements of m with time stamps r. The operation m @+ r is defined at the end of Sect. 5.2. There it is written as m_r, which is more convenient for the formal definitions of Sect. 5.3.

5.2 Timed Multi-sets

In the previous section we have used expressions such as:

$$2`(q,4)@[627,643] + 1`(q,5)@[598]$$

to denote the marking of the places which contain tokens with time stamps. Such an expression indicates that we have two tokens with colour (q,4) and time stamps 627 and 643, while we have one token with colour (q,5) and time stamp 598. More formally, the expression denotes a timed multi-set which we now define. To do this we assume that we have a set of **time values** R, which is a subset of $\mathbb{R}$ closed under + and containing 0. In practice, R will usually be either $\mathbb{R}$ or $\mathbb{Z}$, i.e., the set of all reals or the set of all integers. Timed multi-sets are a modification of ordinary multi-sets (which are defined in Def. 2.1 of Vol. 1).

Definition 5.1: A **timed multi-set** tm, over a non-empty set S, is a function $tm \in [S \times R \rightarrow \mathbb{N}]$ such that the sum:

$$tm(s) = \sum_{r \in R} tm(s,r)$$

is finite for all $s \in S$. The non-negative integer tm(s) is the **number of appearances** of the element s in the timed multi-set tm. The list:

$$tm[s] = [r_1, r_2, \ldots, r_{tm(s)}]$$

is defined to contain the time values $r \in R$ for which $tm(s,r) \neq 0$. Each r appears tm(s,r) times in the list, which is sorted such that $r_i \leq r_{i+1}$ for all $i \in 1..tm(s)-1$.

We usually represent the timed multi-set tm by a formal sum:

$$\sum_{s \in S} tm(s)`s @ tm[s] .$$

By S_{TMS} we denote the set of all timed multi-sets over S. The non-negative integers $\{tm(s) \mid s \in S\}$ are called the **coefficients** of the timed multi-set tm, and tm(s) is called the **coefficient** of s. An element $s \in S$ is said to **belong** to the timed multi-set tm iff $tm(s) \neq 0$ and we then write $s \in tm$.

Each timed multi-set $tm \in S_{TMS}$ determines an ordinary multi-set $tm_U \in S_{MS}$ defined by:

$$tm_U = \sum_{s \in S} tm(s)`s.$$

As an example, consider the timed multi-set:

$$tm = 2`(q,4)@[627,643] + 1`(q,5)@[598]$$

for which we have $tm[(q,4)] = [627,643]$ and $tm[(q,5)] = [598]$, and $tm_U = 2`(q,4) + 1`(q,5)$.

From Def. 5.1 it follows that each timed multi-set over S is also an ordinary multi-set over $S \times R$. This allows us to define +, *, =, ≠, and | |, for timed multi-sets over S, to be identical to the corresponding operations for ordinary multi-sets over S×R; see Def. 2.2 of Vol. 1. As an example, we have:

$$2 * tm = 4`(q,4)@[627,627,643,643] + 2`(q,5)@[598,598]$$

and $|tm| = 3$. We could define ≤ and subtraction of timed multi-sets in a similar way. However, this would *not* be adequate – because $tm_1 \le tm_2$ would then require each element in tm_1 to appear in tm_2 with *exactly* the same time value. This would be too strong a requirement, since we do not want the time stamps of the removed tokens to be identical to the time values specified by the input arc expression. We only require that the time stamps be less than or equal to the specified time values, and hence we give a more liberal definition of ≤. First we define ≤ and subtraction for lists of time values. Then we define ≤ and subtraction for timed multi-sets.

Let $a = [a_1, a_2, \ldots, a_m]$ and $b = [b_1, b_2, .., b_n]$ be two ascending lists over R. We define that $a \le b$ iff $m \le n$ and $a_i \ge b_i$ for all $i \in 1..m$. It should be noted that we require the smallest list to have the largest time values. This reflects the fact that a small time stamp puts fewer restrictions on the use of the token than a large time stamp.

When $a \le b$, we define $b-a$ to be the list, of length $n-m$, which is obtained from b in the following way. From b we remove the largest time value which is smaller than a_1. From the remaining list we remove the largest time value which is smaller than a_2. And so on, until finally, from the remaining list, we remove the largest time value which is smaller than a_m. It can be proved that the result of the list subtraction is independent of the order in which we deal with the elements of a (see Exercise 5.6). For the lists:

$$x = [57, 82, 103, 117, 134, 146]$$
$$y = [98, 136, 145]$$
$$z = [84, 138]$$

we have:

$$(x-y)-z = [57, 103, 146]-z = [146]$$
$$(x-z)-y = [57, 103, 117, 146]-y = [146].$$

It should be noted that we always remove the highest time values – among those which are small enough to be used. Instead it might seem more obvious to remove the smallest time values and define $b-a$ to be the list $[b_{m+1}, b_{m+2}, .., b_n]$. However, we have to choose the first, more complicated solution, since the second, alternative definition would violate a basic Petri net rule, called the diamond rule. This rule tells us that an enabled step with more than one binding element can always be split into two or more substeps that can be executed in any order with the same total effect. The violation of the diamond rule follows from the fact that the alternative definition implies, for the three lists above, that $z \not\le (x-y)$ and $y \not\le (x-z)$ although $(y+z) \le x$, where + denotes concatenation followed by sorting. For the definition which we have chosen, it can be proved that $(y+z) \le x$ implies $(x-y)-z$ and $(x-z)-y$ exist and are equal, for all ascending lists of time values. This property can be generalised to cover the situation where the left-hand side of the inequality contains an arbitrary number of addends (see Exercise 5.6).

Definition 5.2: Comparison of timed multi-sets is defined in the following way, for all $tm_1, tm_2 \in S_{TMS}$:

(i) $tm_1 \leq tm_2 \;=\; \forall s \in S\colon tm_1[s] \leq tm_2[s]$.

When $tm_1 \leq tm_2$ we also define **subtraction**:

(ii) $$tm_2 - tm_1 \;=\; \sum_{s \in S} (tm_2(s) - tm_1(s))`s \;@\, (tm_2[s] - tm_1[s]).$$

It is easy to see that $tm_1[s] \leq tm_2[s]$ implies $tm_1(s) \leq tm_2(s)$. From this it follows that $tm_1 \leq tm_2$ implies $(tm_1)_U \leq (tm_2)_U$, where the first $\leq$ denotes comparison of timed multi-sets over S and the second $\leq$ denotes comparison of ordinary multi-sets over S.

It should be noted that the subtraction operation for timed multi-sets (and for ascending lists of time values) has algebraic properties that are weaker than the subtraction operation for ordinary multi-sets. As an example, we usually do not have $(b-a)+a = b$, but only $((b-a)+a)_U = b_U$.

Our definition of subtraction implies that a token may become "stuck" at a place, because the place may always have other tokens with the same colour and a usable time stamp which is higher. This is a bit awkward, since for tokens with identical colours we might expect a kind of first-in-first-out discipline.

For a timed multi-set $tm \in S_{TMS}$ and a time value $r \in R$ we define $tm_r \in S_{TMS}$ as follows:

$$tm_r \;=\; \sum_{s \in S} tm(s) \;@\, tm[s]_r$$

where $tm[s]_r$ is the list obtained from $tm[s]$ by adding r to each time value. Analogously, for an ordinary multi-set $m \in S_{MS}$ and a time value $r \in R$ we define $m_r \in S_{TMS}$ as follows:

$$m_r \;=\; \sum_{s \in S} m(s) \;\; @\, [r,r,\ldots,r]$$

where the list $[r,r,\ldots,r]$ is of length $m(s)$ for all $s \in S$.

5.3 Formal Definition of Timed CP-nets

We are now ready to give the formal definition of timed CP-nets.

Definition 5.3: A **timed** non-hierarchical CP-net is a tuple TCPN = (CPN, R, r_0) such that:

(i) CPN satisfies the requirements of a non-hierarchical CP-net in Def. 2.5 of Vol. 1 – when in (viii) and (ix) we allow the type of E(a) and I(p) to be a timed *or* an untimed multi-set over C(p(a)) and C(p), respectively.
(ii) R is a set of **time values**, also called **time stamps**. It is a subset of $\mathbb{R}$ closed under + and containing 0.
(iii) r_0 is an element of R, called the **start time**.

Timed CPN models often contain one or more colour sets S which are **untimed**. This means the tokens of type S are required to be always available, independently of any time constraints. An example is the colour set E in Fig. 5.1. An untimed colour set has initialization expressions and arc expressions of type S_{MS}. This guarantees that the time stamps will always be small enough (see the enabling and occurrence rule defined below).

The formal definition of timed CP-nets allows the use of negative time values in arc expressions and initialization expressions. It is not obvious that this is useful and it can easily be ruled out.

The set of bindings B(t), token elements TE, binding elements BE and steps $\mathbb{Y}$ are defined in exactly the same way as for an untimed CP-net; see Defs. 2.6–2.7 of Vol. 1.

Definition 5.4: A **marking** is a timed multi-set over TE. The **initial marking** M_0 is the marking obtained by evaluating the initialization expressions:

$$\forall p \in P: M_0(p) = I(p)_{r_0}.$$

A **state** is a pair (M, r) where M is a marking and r a time value. The **initial state** is the pair (M_0, r_0).

The sets of all markings and states are denoted by $\mathbb{M}$ and $\mathbb{S}$, respectively.

For a state $S = (M,r)$, we use S_U to denote the untimed multi-set $M_U \in TE_{MS}$. S_U is called the **untimed marking** determined by S. Intuitively, it is obtained by discarding all the time information from the state S. For a set of states $X \subseteq \mathbb{S}$, we use X_U to denote the set of all untimed markings determined from states in X:

$$X_U = \{M \in TE_{MS} \mid \exists S \in X: M = S_U\}.$$

Definition 5.5: A step Y is **enabled** in a state (M_1, r_1) at time r_2 iff the following properties are satisfied:

(i) $\forall p \in P: \sum_{(t,b) \in Y} E(p,t)\langle b\rangle_{r_2} \leq M_1(p)$.

(ii) $r_1 \leq r_2$.

(iii) r_2 is the smallest element of R for which there exists a step satisfying (i) and (ii).

From the remarks immediately below Def. 5.2 we conclude that (i) implies:

$$\forall p \in P: \sum_{(t,b) \in Y} (E(p,t)\langle b\rangle_{r_2})_U \leq (M(p))_U$$

which is the standard enabling rule for the untimed CP-net obtained from TCPN by discarding all time information, i.e., replacing each timed multi-set tm by the ordinary multi-set tm_U.

The first condition of Def. 5.5 guarantees that we only execute steps for which we have the necessary tokens with time stamps which are small enough, i.e., steps which are colour enabled and ready. The second condition guarantees that time cannot go backwards. Finally, the third condition guarantees that colour enabled steps are executed in the order in which they become ready. It should be noted that the definitions of tm_r and m_r (at the end of Sect. 5.2) imply that the enabling rule in (i) works for both timed and untimed arc expressions (i.e., arc expressions of type $C(p)_{TMS}$ and of type $C(p)_{MS}$). The same is true for the occurrence rule defined below.

Definition 5.5 (iii) implies that the enabling rule is no longer local. To determine whether a transition has an enabled binding element at time r_2, we need to consider all transitions that have colour enabled binding elements. Otherwise we cannot know whether r_2 is minimal.

When we have infinite colour sets and a dense set of time values R, we may have a marking in which Def. 5.5 (iii) cannot be satisfied (because the time values at which the steps become ready form an infinite set with no minimum). This means the marking is dead. A timed CP-net where this can happen is said to be **blocking**.

We define **enabled** transitions and **concurrently enabled** transitions/binding elements analogously to the corresponding concepts in an untimed CP-net, see Def. 2.8 of. Vol. 1.

Definition 5.6: When a step Y is enabled in a state (M_1, r_1) at time r_2 it may **occur**, changing the state (M_1, r_1) to another state (M_2, r_2), where M_2 is defined by:

$$\forall p \in P: M_2(p) = \Big(M_1(p) - \sum_{(t,b)\in Y} E(p,t)\langle b\rangle_{r_2}\Big) + \sum_{(t,b)\in Y} E(t,p)\langle b\rangle_{r_2}.$$

The first sum is called the **removed tokens** while the second is called the **added tokens**. Moreover, we say that (M_2, r_2) is **directly reachable** from (M_1, r_1) by the occurrence of the step Y at time r_2, which we also denote: $(M_1, r_1)\,[Y, r_2\rangle\,(M_2, r_2)$.

The occurrence rule in Def. 5.6 is identical to the occurrence rule for untimed CP-nets, except for the two appearances of r_2. It should be noted that the occurrence rule is local, even though the enabling rule is not. The following definitions of occurrence sequences and reachability are identical to the corresponding definitions for untimed nets (see Defs. 2.10 and 2.11 of Vol. 1) – except that time values are added.

Definition 5.7: A **finite occurrence sequence** is a sequence of states, steps and time values:

$$S_1\,[Y_1, r_2\rangle\, S_2\,[Y_2, r_3\rangle\, S_3 \ldots S_n\,[Y_n, r_{n+1}\rangle\, S_{n+1}$$

such that $n \in \mathbb{N}$, and $S_i\,[Y_i, r_{i+1}\rangle\, S_{i+1}$ for all $i \in 1..n$. The state S_1 is called the **start state** of the occurrence sequence, while the state S_{n+1} is called the **end state**. The non-negative integer n is the **number of steps** in the occurrence sequence, or the **length** of it.

Analogously, an **infinite occurrence sequence** is a sequence of markings, steps and time values:

$$S_1\,[Y_1, r_2\rangle\, S_2\,[Y_2, r_3\rangle\, S_3 \ldots$$

such that $S_i\,[Y_i, r_{i+1}\rangle\, S_{i+1}$ for all $i \in \mathbb{N}_+$. The state S_1 is called the **start state** of the occurrence sequence, which is said to have **infinite length**.

The set of all finite occurrence sequences is denoted by OSF, while the set of all infinite occurrence sequences is denoted by OSI. Moreover, we use $OS = OSF \cup OSI$ to denote the set of all occurrence sequences.

A state S" is **reachable** from a state S' iff there exists a finite occurrence sequence having S' as start state and S" as end state. As a shorthand, we say that S is reachable iff it is reachable from S_0. The *set* of states which are reachable from the state S is denoted $[S\rangle$.

We may omit certain parts of a timed occurrence sequence in a similar way as for untimed CP-nets. In addition we often omit some of the time values – in the states, in the steps, or both.

We have shown above how to define timed non-hierarchical CP-nets. It should be obvious that the same method can be used to obtain timed hierarchical CP-nets. The formal definitions are straightforward, and hence they are omitted.

Next we consider the definitions of the dynamic properties from Chap. 4 of Vol. 1. With very small modifications these definitions can be used for timed CP-nets also.

For the boundedness properties in Defs. 4.6 and 4.7 of Vol. 1, we replace $[M_0\rangle$ by $[S_0\rangle_U$. Otherwise, there are no modifications.

For the home properties in Def. 4.8 of Vol. 1, we replace the markings M_0 and M' by states S_0 and S'. An untimed marking $M \in \mathbb{S}_U$ is a **home marking** iff:

$$\forall S' \in [S_0\rangle: M \in [S'\rangle_U.$$

A set of untimed markings $X \subseteq \mathbb{S}_U$ is a **home space** iff:

$$\forall S' \in [S_0\rangle: X \cap [S'\rangle_U \neq \emptyset.$$

For the liveness properties in Def. 4.10 of Vol. 1, we replace the markings M, M', and M" by states S, S', and S". Otherwise there are no modifications and Prop. 4.11 of Vol. 1 is still satisfied.

For the fairness properties in Defs. 4.12 and 4.14 of Vol. 1, $EN_X(\sigma)$ and $OC_X(\sigma)$ deal with occurrence sequences which contain states instead of markings. Otherwise there are no modifications and Prop. 4.13 of Vol. 1 is still satisfied.

5.4 Relationships Between Timed and Untimed CP-nets

From the definitions in Sects. 5.2 and 5.3, it is straightforward to see that we can transform a timed CP-net into an untimed CP-net simply by discarding all the time information in the arc expressions and the initialization expressions:

Definition 5.8: Each timed CP-net TCPN = (CPN, R, r_0) **determines** an untimed CP-net UCPN, which we obtain from CPN by replacing each timed arc expression E(a) and each timed initialization expression I(p) by $E(a)_U$ and $I(p)_U$, respectively.

Each finite occurrence sequence of TCPN:

$$S_1 [Y_1, r_2\rangle S_2 [Y_2, r_3\rangle S_3 \ldots S_n [Y_n, r_{n+1}\rangle S_{n+1}$$

determines a finite occurrence sequence of UCPN defined by:

$$(S_1)_U [Y_1\rangle (S_2)_U [Y_2\rangle (S_3)_U \ldots (S_n)_U [Y_n\rangle (S_{n+1})_U.$$

A similar property is satisfied for infinite occurrence sequences.

For the second part of Def. 5.8 to make sense, it is necessary to prove that:

$$(S_1)_U [Y_1\rangle (S_2)_U [Y_2\rangle (S_3)_U \ldots (S_n)_U [Y_n\rangle (S_{n+1})_U$$

is an occurrence sequence of UCPN. This is a straightforward consequence of the enabling and occurrence rules of timed and untimed CP-nets, and hence the proof is omitted.

It should be noted that UCPN often has occurrence sequences which cannot be determined from an occurrence sequence of TCPN. This is because the time constraints of TCPN limit the possibilities, rather like guards and input arc expressions. As an example, let us again consider the resource allocation system (without cycle counters). For the untimed CP-net in Fig. 1.1 we have the O-graph shown in Fig. 5.3 (except for the line thickness, it is identical to Fig. 1.2). For the timed CP-net, it is only possible to reach some of the markings in Fig. 5.3. When the timed CP-net reaches a marking represented by node #2, it will only be the binding element (T1,Q) which is enabled. The binding element (T3,P) is colour enabled, but it is not ready because it needs a token which was created in the previous step – with a time stamp 8 time units larger than the current model time. A similar property is true for node #10. From this, it is easy to deduce that the timed CP-net – after the initial one or two steps – only can traverse the highlighted part of Fig. 5.3. Hence it is only in node #5 that we have a conflict, i.e., a choice between several binding elements. Otherwise, the behaviour of the timed CP-net is totally deterministic.

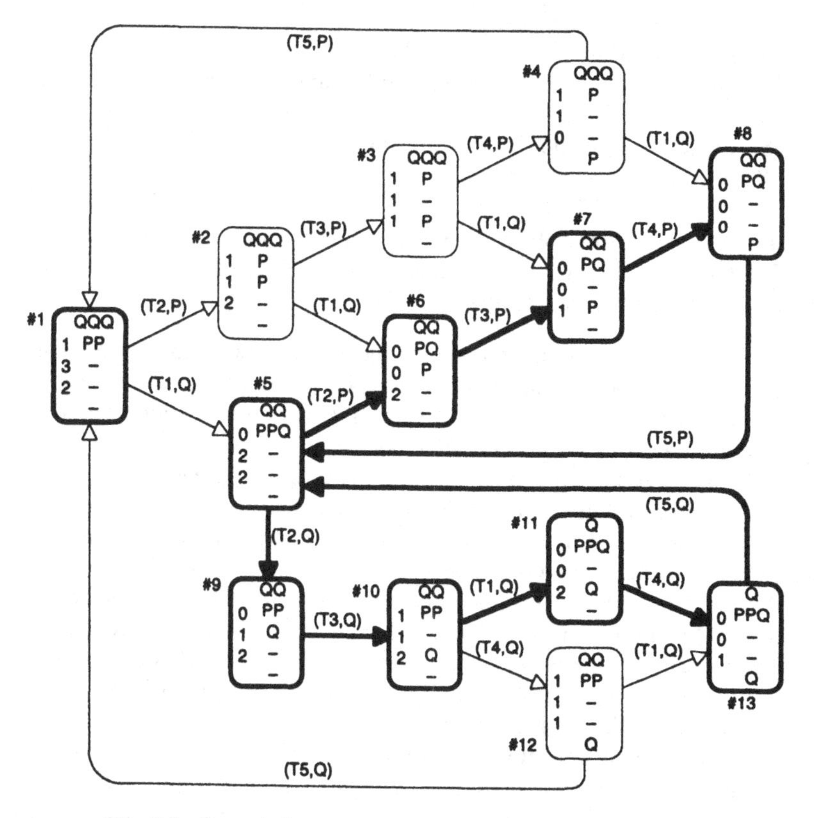

Fig. 5.3. O-graph for the resource allocation system in Fig. 1.1

The above example indicates that it is important to choose the time delays in a suitable way. Unfortunately, a relatively small change in the time delays may imply that the behaviour of a timed CP-net is dramatically changed. In order to reduce such problems, it is a good idea to repeat the simulations with different time delays. Moreover, it is advisable to specify the delays by means of probability distributions instead of fixed time values. One of the simplest solution is to let the delay of a transition $t \in T$ belong to an interval tMin..tMax, with equal probability for all values in the interval. In Sect. 5.5 we show how such a delay can be specified. It is easy to generalise the method to cover more complex probability distributions.

We have the following relationship between the dynamic properties of a timed CP-net and the dynamic properties of the corresponding untimed CP-net. The left-hand side of the implications are properties of the untimed net, while the right-hand sides are properties of the timed net.

Proposition 5.9: Let UCPN be the untimed CP-net determined from a timed CP-net TCPN. We then have the following properties for all states S, $S_1, S_2 \in \mathbb{S}$, all untimed multi-sets m, all integers $n \in \mathbb{N}$, and all sets of binding elements $X \subseteq BE$:

(i) $(S_2)_U \in [(S_1)_U\rangle \Leftarrow S_2 \in [S_1\rangle$.
(ii) n valid bound in UCPN $\Rightarrow$ n valid bound in TCPN.
(iii) m valid bound in UCPN $\Rightarrow$ m valid bound in TCPN.
(iv) S_U dead $\Rightarrow$ S dead.
(v) S_U dead $\Leftarrow$ S dead $\wedge$ TCPN non-blocking.
(vi) X dead in S_U $\Rightarrow$ X dead in S.

Proof: All properties follow from the fact that each occurrence sequence of TCPN determines an occurrence sequence of UCPN. The proofs are straightforward and hence they are omitted. □

Properties (ii) and (iii) tell us that any bound (upper/lower, integer/multi-set, set of token elements/place instance) which is valid for the untimed net also is valid for the timed net (as the same kind of bound). However, it is not guaranteed that a given bound remains the best possible bound.

For home markings, liveness, and fairness it is – *in general* – impossible to deduce any information about the timed CP-net from the untimed net, or vice versa. Nevertheless, in practice we can often obtain such information. To illustrate this, let us assume that we expect the marking M_H to be a home marking of TCPN but discover that M_H is not a home marking of UCPN because there exists a marking M from which we cannot reach M_H in UCPN. From Prop. 5.9 (i) we then know that we cannot reach M_H from M in TCPN. If we can show that $M \in [S_0\rangle_U$, it follows that M_H cannot be a home marking of TCPN. Similar arguments can be made for liveness and fairness properties.

For many kinds of system it is possible to specify adequate behaviour without making any specific time assumptions. This means the system should work correctly with all kinds of time delays. Then it is reasonable to analyse the dynamic properties by means of the untimed CP-net. It is only when we turn to the task of optimising the performance of such systems that it becomes necessary to introduce time assumptions, i.e., analyse the timed CP-net.

5.5 Example of Timed CP-nets: Protocol

This section contains a second example of timed CP-nets. We describe a simple protocol where a sequence of packets is sent from one site to another via a network where packets may be delayed or lost. The purpose of our description is to show how timed nets can be used to model and evaluate protocols, in particular their performance. We do not claim that the described protocol is optimal. However, the protocol is interesting enough to deserve a closer investigation, and it is also complex enough for such an investigation to be necessary.

The protocol system is shown in Fig. 5.4. Again, we use a discrete clock, starting at 0. The protocol system consists of three parts. The *Sender* part has two transitions which can *Send Packets* and *Receive Acknowledgements*. The *Network* part has two transitions which can *Transmit Packets* and *Transmit Acknowledgements*. Finally, the *Receiver* part has a single transition which can *Receive Packets* (and send acknowledgements). The interface between the *Sender* and the *Network* consists of the places *A* and *D*, while the interface between the *Network* and the *Receiver* consists of the places *B* and *C*.

The packets to be sent are positioned at the place *Send* (in the upper left corner). Each token on this place consists of a packet number and the data contents of the packet (which we represent as a text string). The place *Next Send* indicates the number of the next packet to be sent. Initially this number is 1, and it is updated each time an acknowledgement is received.

The packets that have been received are positioned at the place *Received* (in the upper right corner). The tokens on this place have the same format as the tokens on *Send*. The place *Next Rec* indicates the number of the next packet to be received. Initially this number is 1, and it is updated each time a packet is successfully received.

At the end of the transmission we expect *Send* and *Received* to have the same marking, when time stamps are ignored. This means all packets are received, without duplication. Furthermore, we expect the tokens on *Received* to have time stamps that increase with the packet number. This means the packets are received in the same order as they are sent. The above properties can be verified from the final marking shown in Fig. 5.5. However, it is necessary to simulate the timed CP-net with a much larger number of packets and with different delay values to become convinced that the protocol works. Alternatively, we can investigate the protocol system by the more formal analysis methods discussed in Sect. 5.6.

We do not model how the *Sender* splits a message into a sequence of packets or how the *Receiver* reassembles the packets into a message. Neither do we model how the tokens at *Send* and *Received* are removed at the end of the transmission or how the packet numbers in *Next Send* and *Next Rec* are reset to 1. Now let us take a closer look at the five different transitions in the protocol system.

Send Packet sends a packet to the *Network* by creating a copy of the packet on the place *A*. The number in *Next Send* specifies which packet to send. The transition has a time inscription: $@+T_{sp}$. This is a shorthand that implies that a common delay T_{sp} is added to the time stamps of *all* output tokens. The tokens created at *A* and *Next Send* get a time stamp that is the current time r^* (at which the transition occurs) plus T_{sp}. The token created for *Send* gets a time stamp $r^*+T_{sp}+T_{wait}$. Intuitively, the delay T_{sp} represents the time used to send a packet, while the delay T_{wait} represents the time that has to elapse before a retransmission, i.e., before *Send Packet* occurs once more for the same packet.

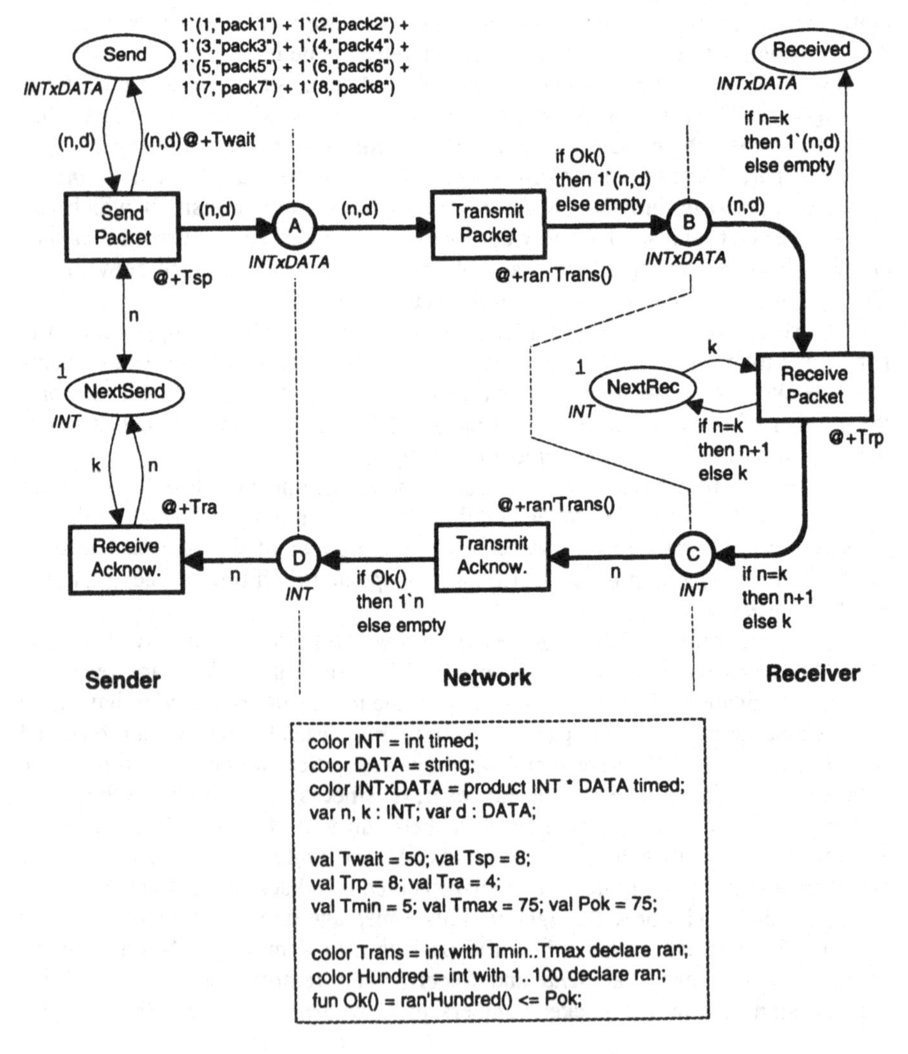

Fig. 5.4. Timed CP-net for the protocol system

A retransmission will only happen if the number in *Next Send* remains unaltered for $T_{sp}+T_{wait}$ time units, i.e., if no acknowledgement is received inside this time period. From the middle part of the declarations, we see that $T_{sp} = 8$ and $T_{wait} = 50$.

Transmit Packet transmits a packet from the *Sender* site of the *Network* to the *Receiver* site by moving the corresponding token from *A* to *B*. The duration of this operation is determined by the function *ran'Trans()*, which returns a random number between T_{min} and T_{max}. It is a predefined function which is available due to the declare clause in the declaration of the *Trans* colour set. The CPN simulator described in Chap. 6 of Vol. 1 offers a variety of predeclared functions implementing different kinds of time delays (probability distributions). For an explanation of predefined functions see the end of Sect. 1.3 of Vol. 1. The function *Ok()* in the output arc expression determines whether the packet is successfully transmitted or lost. The probability for successful transmission is determined by the constant P_{ok}.

Receive Packet receives a packet and checks whether the packet number is identical to the number in *Next Rec*. When the two numbers match, the packet is moved to *Received* and the number in *Next Rec* is increased by 1. Otherwise, the packet is ignored and the number in *Next Rec* is left unchanged. In both cases an

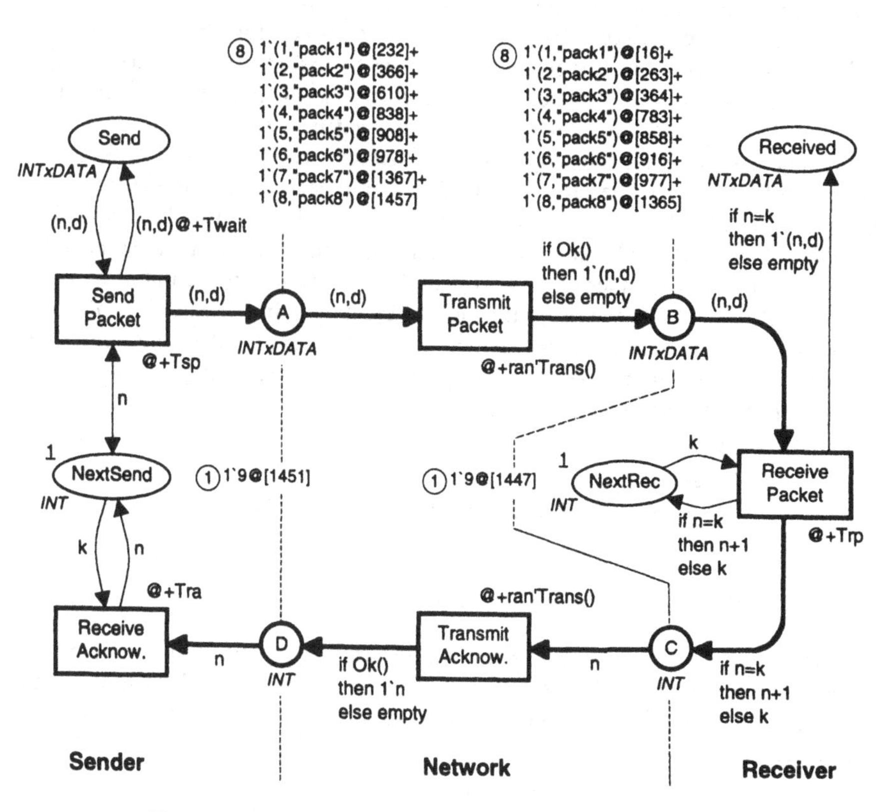

Fig. 5.5. Timed CP-net for the protocol system (final marking)

acknowledgement is sent containing the number of the next packet which the *Sender* should transmit. The duration of the operation is determined by the constant T_{rp}.

Transmit Acknowledgement transmits an acknowledgement from the *Receiver* site of the *Network* to the *Sender* site by moving the corresponding token from *C* to *D*. The transition works in a similar way as *Transmit Packet.* This means the transmission delay is randomly determined, between T_{min} and T_{max}. It also means the acknowledgement may be lost, with a probability determined by P_{ok}.

Receive Acknowledgement receives an acknowledgement and updates the number in *Next Send.* The duration of the operation is determined by the constant T_{ra}.

Note that the token in *Next Send* has a time stamp. Intuitively, this means we cannot start a new *Send Packet* or a new *Receive Acknowledgement* as long as one of these operations is already ongoing. If the *Sender* has multiple processes (threads), allowing an unlimited number of *Sender* operations to be performed at the same time, we simply make the colour set of *Next Send* untimed. A similar remark applies for the operations of the *Receiver* and the colour set of *Next Rec.*

By means of our timed CPN model we can investigate whether the protocol works correctly, e.g., whether packets are received in the correct order, without loss and without duplication. Furthermore, we can investigate the performance of the protocol, e.g., experiment with different values for the retransmission delay T_{wait}. A short delay increases the chance of making unnecessary retransmissions and hence the chance that a *Receive Acknowledgement* operation is postponed, because the *Sender* process is engaged in a retransmission. A long delay means it may take too long before the *Sender* recognises that a packet or an acknowledgement has been lost. By making a number of simulations, with different values for T_{wait}, we can determine the optimal value for the retransmission delay. To obtain reliable results we must of course perform the simulations with a much larger number of packets than the 8 indicated in the initial marking of Fig. 5.4. To do this we use a function to create the initial marking of *Send.* This function can be declared as shown below. The argument n determines the number of packets to be created. The predeclared function *makestring* returns the string representation of the integer n.

```
fun Packets(n:int) =
        if n=0 then empty
                else 1`(n,"pack"^makestring(n))+Packets(n-1).
```

5.6 Analysis of Timed CP-nets

In this section we discuss how simulation, occurrence graphs, and invariants can be used to analyse timed CP-nets. To a very large extent, this is done in the same way as for untimed CP-nets. Hence we only describe the necessary modifications and extensions.

Simulation

Simulation of timed CP-nets is performed in the same way as for untimed nets, except that the simulator now uses the enabling rule and the occurrence rule of timed nets.

The CPN simulator described in Chap. 6 of Vol. 1 supports simulation of timed CP-nets. The current version of the simulator only allows time values in output arc expressions. Future versions will support a larger subset of the possibilities described in this chapter.

The simulation facilities that can be used for timed CP-nets are the same as those that can be used for untimed CP-nets. In particular this means a timed simulation can be manual, automatic, or super-automatic, and with or without code. For a timed CP-net, it is also possible to determine whether the simulation is with or without time. In the latter case the simulator uses the untimed CP-net determined by the timed net.

A timed simulation is often done to obtain different performance measures of the modelled system. To support this, the CPN simulator has facilities that make it easy to measure the flow of tokens through a given place, e.g., the average number of tokens and the average waiting time. It is often convenient to display such information in charts. As an example, see the chart in Fig. 5.6. The chart shows the performance of the timed resource allocation system from Fig. 5.1, by displaying the factors by which p-processes and q-processes are delayed, compared to the speed the processes would have if they did not have to wait for resources. A chart may be updated during the simulation – with a specified period. In this way the user can follow the results of the simulation without investigating the individual states. He can also postpone all chart updating to the end of the simulation. This will increase the speed.

As mentioned in Sect. 5.3, timed CP-nets have a non-local enabling rule. This creates a complexity problem for the simulation of large timed CP-nets, because the calculation of the enabling grows with the number of transition instances, instead of being independent of the size of the model (see the discussion of "simulation of large CPN diagrams" in Sect. 6.2 of Vol. 1). One way to over-

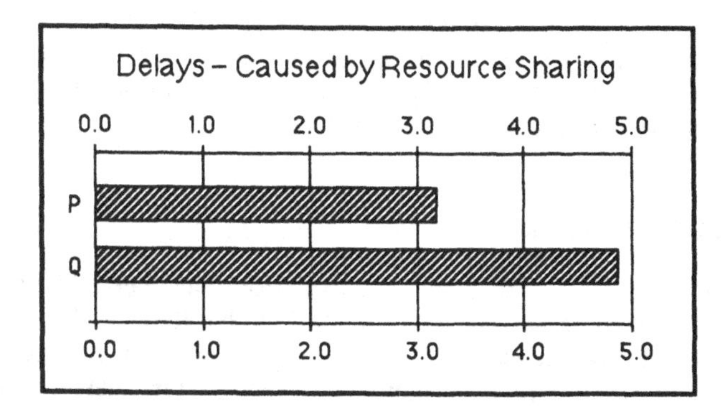

Fig. 5.6. Chart with performance measures of the resource allocation system

come this complexity problem might be to replace the single global clock by a set of local clocks, e.g., one for each page instance. Such an approach would also have interesting theoretical implications. A discussion of local clocks is outside the scope of this book.

Occurrence graphs

Occurrence graphs of timed CP-nets are defined in the same way as for untimed nets, except that the nodes now represent states instead of markings. This means each node contains a time value and a timed marking.

For a non-cyclic CP-net, a timed occurrence graph will often be much smaller than the corresponding untimed occurrence graph, because the time constraints limit the possible orders in which binding elements may occur. However, if we use non-deterministic time delays, with a large number of possible time values, we may get a larger occurrence graph.

For a cyclic system, a timed occurrence graph usually become infinite, because each repeated appearance of a marking corresponds to a new state and hence implies the creation of a new node. For occurrence graphs with equivalence classes, we may use the equivalence relation to avoid this explosion in the number of nodes. Otherwise, it may become necessary to construct a partial graph, e.g., the subgraph that contains all those states which are created before a certain moment of time. Such a subgraph, although partial, may be very useful to determine the dynamic properties and the performance characteristics of the modelled system.

Invariants

Unfortunately, it is not easy to modify the invariant method so that it becomes applicable to prove properties about the time stamps of a timed CP-net. The main problem is the fact that the sets of removed tokens are not fully determined by the binding elements, because we only require the time stamps to be small enough, instead of requiring them to have some exact time values. This means linearity of weight functions (between timed multi-sets) is insufficient to guarantee that each flow determines an invariant.

It is of course possible to use invariants to analyse the untimed CP-net determined from a timed CP-net. In this way it is possible to prove dynamic properties that are independent of the time constraints.

Bibliographical Remarks

A considerable number of different time extensions have been defined for Petri nets, and many of these have been generalised so that they can be used also for CP-nets. Most of the time extensions use a global clock that may be continuous or discrete. However, the time consumption is described in many different ways. Some time extensions enforce a delay between the enabling and the occurrence of a transition. Others have a delay between the removal of input tokens and the creation of output tokens, or a delay between the creation of a token and the time at which that token can be used. The latter method is very close to the timed CP-nets defined in this book, because the time stamps usually make new tokens unavailable for some period of time. The idea of using time stamps is due to the group behind the ExSpect tool developed at Eindhoven University of Technology. For more information see [1] and [20].

It has been shown that most of the time extensions mentioned above have the same modelling power – in the sense that it is relatively easy to translate a Petri net model using one of the time extensions into a model that uses another kind of time extension. We have chosen to attach time stamps to the individual tokens, because it is a rather natural and straightforward approach – in particular for high-level nets, where each token already carries a token colour. As shown in Sect. 5.3, the formal definitions of timed CP-nets and their behaviour become straightforward modifications of the corresponding definitions for untimed CP-nets. Moreover, we get a simple relationship between the behaviour of a timed CP-net and the behaviour of the corresponding untimed CP-net, as illustrated by the second half of Def. 5.8.

It is also important to consider the various kinds of delays that the different time extensions allow. Some time extensions only allow constant delays. Others allow random delays within intervals, or delays which are determined by probability distributions, e.g., negative exponential functions. The timed CP-nets and the CPN tools described in this book allow all kinds of delays. This is due to the fact that we do not impose any restrictions on the timed multi-sets to which the arc expressions evaluate.

For certain kinds of delays, it is possible to translate a timed CP-net into a Markov chain (which is a well known type of statistical model). The Markov chain determines an equation system, by means of which we can find analytic solutions to the different performance measures. The solutions are found directly from the equation system and thus they are general (while results found via simulation always depend, at least to some extent, upon the chosen occurrence sequences). For some kinds of CP-nets it is faster (in terms of CPU time) to obtain analytic solutions – compared to the execution of lengthy simulation runs. However, many CP-nets are too complex to be analysed via Markov chains (because the equation system has too many unknown variables and thus becomes too complex to solve).

Performance analysis is one of the largest and most important subareas of Petri nets. Different time extensions of CP-nets are described in [4], [7], [15], [29], and in many of the papers contained in [35], [36], and [37].

Two examples of the practical use of the timed CP-nets presented in this chapter can be found in [5] and [12]. Both of these have obtained interesting, non-trivial performance results by means of the CPN simulator described in [24] and Chap. 6 of Vol. 1.

Exercises

Exercises 5.1–5.5 are only worth attempting if you have access to a simulation tool and/or an occurrence graph tool for timed CP-nets.

Exercise 5.1.
Consider the resource allocation system from Fig. 5.1.

(a) Make a simulation of the timed CP-net and investigate the factors by which p-processes and q-processes are delayed, compared to the speed the processes would have if they did not have to wait for resources. Compare the results with the bar chart in Fig. 5.6.

(b) Construct a timed occurrence graph for the resource allocation system and use it to deduce information about the performance measures described in (a). Compare the results to those which you obtained by means of simulation.

Exercise 5.2.
Consider the protocol system from Fig. 5.4.

(a) Investigate how long time it takes to transmit 100 packets, for different values of T_{wait}. What is the optimal value?

(b) Construct a number of timed occurrence graphs for the protocol system and use them to deduce information about the performance measure described in (a). Compare the results to those which you obtained by means of simulation.

Exercise 5.3.
Consider the ring network from Sect. 3.1 of Vol. 1.

(a) Introduce time delays such that:
   - The delay for *NewPack* is determined by a random number in an interval $T_{n1}..T_{n2}$.
   - The delay for *Send* is a constant T_S.
   - The delay for *Receive* is determined by a random number in an interval $T_{r1}..T_{r2}$.

(b) Make a simulation of the timed CP-net and determine the average number of packets on the network (i.e., on *1to2*, *2to3*, *3to4*, and *4to1*) and the average number of packets which are waiting on *Packet*.

(c) Construct a timed occurrence graph for the ring network and use it to deduce information about the performance measures described in (b). Compare the results to those which you obtained by means of simulation.

Exercise 5.4.
Consider an elevator system with two separate shafts. The two elevators serve five different floors. Each floor has an UP and a DOWN button (the upper floor only has DOWN, while the lower floor only has UP). All buttons are shared by the two shafts.

(a) Construct a CP-net modelling the elevator system. This includes the formulation of a service strategy describing the order in which requests are honoured.

(b) Modify your CPN model so that it becomes a timed CP-net with suitable delays for the various operations, such as the movement between floors and the opening and closing of doors.

(c) Make a number of timed simulations to investigate the performance of the elevator system for different work loads. The elevator requests may be generated by means of transitions or they may be defined via the initial marking.

(d) Investigate whether you can modify your service strategy so that the performance of the elevator system is improved.

Exercise 5.5.
Consider the telephone system from Sect. 3.2 of Vol. 1.

(a) Modify the CP-net so that the conflicts for the *RepXXX* transitions are resolved as follows:

- RepCont / Dial = 1 / 9.
- RepNoTone / Engaged = 1 / 7.
- RepLong / LiftRinging = 1 / 5.
- RepConSen / RepConRec = 1 / 1.
- RepRepl / LiftRepl = 4 / 1.

This means *RepCont* in average occurs once whenever *Dial* occurs 9 times, and so on.

(b) Introduce some sensible time delays for the individual transitions.

(c) Make a simulation of the timed CP-net and determine the average number of *Requests*, *Calls*, and *Connections*.

(d) Construct a timed occurrence graph for the telephone system and use it to deduce information about the performance measures described in (c). Compare the results to those which you obtained by means of simulation.

Exercise 5.6.
In this exercise we consider subtraction of lists and subtraction of timed multi-sets.

(a) Prove that the result of the list subtraction b – a is independent of the order in which we deal with the elements of a.

(b) Use (a) to prove that the inequality $(y+z) \leq x$ implies that $(x-y)-z$ and $(x-z)-y$ exist and are equal, for all ascending lists of time values.

(c) Prove that the property in (b) can be generalised to cover the situation where the left hand-side of the inequality contains an arbitrary number of addends.

(d) Prove that the results in (b) and (c) remain valid when x, y and z are timed multi-sets over a set S.

(e) Prove that each enabled step with more than one binding element can always be split into two or more substeps which can be executed in any order with the same total effect.

Exercise 5.7.
Consider Prop. 5.9 which deals with the relationship between the dynamic properties of a timed CP-net and the dynamic properties of the corresponding untimed CP-net.

(a) Fill in the details of the proofs for properties (i)–(vi).

References

1. W.M.P. van der Aalst: *Interval Timed Coloured Petri Nets and Their Analysis*. In: M. Ajmone Marsan (ed.): Application and Theory of Petri Nets 1993. Proceedings of the 14th International Petri Net Conference, Chicago 1993, Lecture Notes in Computer Science Vol. 691, Springer-Verlag 1993, 453–472.
2. A.V. Aho, J.E. Hopcroft, J.D. Ullman: *The Design and Analysis of Computer Algorithms*. Addison-Wesley, 1974.
3. M. Beaudouin-Lafon: *The Graph Widget – User's Manual*. Technical Report, Université de Paris-Sud, Laboratoire de Recherche en Informatique, 1991.
4. J.A. Carrasco: *Automated Construction of Compound Markov Chains from Generalized Stochastic High-level Petri Nets*. In [35], 93–102. Also in [26], 494–503.
5. L. Cherkasova, V. Kotov, T. Rokicki: *On Scalable Net Modeling of OLTP*. In [37], 270-279.
6. G. Chiola, C. Dutheillet, G. Franceschinis, S. Haddad: *On Well-Formed Coloured Nets and Their Symbolic Reachability Graph*. In [26], 373–396.
7. G. Chiola, C. Dutheillet, G. Franceschinis, S. Haddad: *Stochastic Well-Formed Coloured Nets and Multiprocessor Modelling Applications*. In [26], 504–530.
8. C. Chiola, C. Dutheillet, G. Franceschinis, S. Haddad: *A Symbolic Reachability Graph for Coloured Petri Nets*. To appear in Theoretical Computer Science, North-Holland.
9. S. Christensen, L. Petrucci: *Towards a Modular Analysis of Coloured Petri Nets*. In: K. Jensen (ed.): Application and Theory of Petri Nets 1992. Proceedings of the 13th International Petri Net Conference, Sheffield 1992, Lecture Notes in Computer Science Vol. 616, Springer-Verlag 1992, 113–133.
10. S. Christensen, J. Toksvig: *Tool Support for Place Flow Analysis of Hierarchical CP-nets*. Computer Science Department, Aarhus University, Denmark, 1993.
11. E.M. Clarke, T. Filkorn, S. Jha: *Exploiting Symmetry in Temporal Logic Model Checking*. In: C. Courcoubetis (ed.): Computer Aided Verification. Proceedings of the 5th International Conference on Computer Aided Verification, Elounda, Greece, 1993, Lecture Notes in Computer Science Vol. 697, Springer-Verlag 1993, 450–462.

12. H. Clausen, P.R. Jensen: *Validation and Performance Analysis of Network Algorithms by Coloured Petri Nets.* In [37], 280–289.

13. J.M. Couvreur, S. Haddad, J.F. Peyre: *Generative Families of Positive Invariants in Coloured Nets Sub-Classes.* In: G. Rozenberg (ed.): Advances in Petri Nets 1993, Lecture Notes in Computer Science Vol. 674, Springer-Verlag 1993, 51–70.

14. J.M. Couvreur, J. Martínez: *Linear Invariants in Commutative High Level Nets.* In: G. Rozenberg (ed.): Advances in Petri Nets 1990, Lecture Notes in Computer Science Vol. 483, Springer-Verlag 1991, 146–165. Also in [26], 284–302.

15. C. Dutheillet, S. Haddad: *Regular Stochastic Petri Nets.* In: G. Rozenberg (ed.): Advances in Petri Nets 1990, Lecture Notes in Computer Science Vol. 483, Springer-Verlag 1991, 186–210. Also in [26], 470–493.

16. E.A. Emerson, A.P. Sistla: *Symmetry and Model Checking.* In: C. Courcoubetis (ed.): Computer Aided Verification. Proceedings of the 5th International Conference on Computer Aided Verification, Elounda, Greece, 1993, Lecture Notes in Computer Science Vol. 697, Springer-Verlag 1993, 463–477.

17. A. Field, P. Harrison: *Functional Programming.* International Computer Science Series. Addison-Wesley, 1988.

18. A. Finkel: *The Minimal Coverability Graph for Petri Nets.* In: G. Rozenberg (ed.): Advances in Petri Nets 1993, Lecture Notes in Computer Science Vol. 674, Springer-Verlag 1993, 210–234.

19. A. Gibbons: *Algorithmic Graph Theory.* Cambridge University Press, 1985.

20. K.M. van Hee, L.J. Somers, M. Voorhoeve: *Executable Specifications for Distributed Information Systems.* In: E.D. Falkenberg, P. Lindgreen (eds.): Proceedings of the IFIP TC8/WG 8.1 Working Conference on Information System Concepts, Namur, Belgium 1989, Elsevier Science Publishers, 1989, 139–156.

21. P. Huber, A.M. Jensen, L.O. Jepsen, K. Jensen: *Reachability Trees for High-level Petri Nets.* Theoretical Computer Science 45 (1986), North-Holland, 261–292. Also in [26], 319–350.

22. K. Jensen: *Coloured Petri Nets and the Invariant Method.* Theoretical Computer Science 14 (1981), North-Holland, 317–336.

23. K. Jensen: *Coloured Petri Nets.* In: W. Brauer, W. Reisig, G. Rozenberg (eds.): Petri Nets: Central Models and Their Properties, Advances in Petri Nets 1986 Part I, Lecture Notes in Computer Science Vol. 254, Springer-Verlag 1987, 248–299.

24. K. Jensen, et. al: *Design/CPN. A Reference Manual.* Meta Software and Computer Science Department, University of Aarhus, Denmark. On-line version: http://www.daimi.aau.dk/designCPN/.

25. K. Jensen, et. al.: *Design/CPN Occurrence Graph Manual.* Meta Software and Computer Science Department, University of Aarhus, Denmark. On-line version: http://www.daimi.aau.dk/designCPN/.

26. K. Jensen, G. Rozenberg (eds.): *High-level Petri Nets. Theory and Application.* Springer-Verlag, 1991.

27. T. Kamada, S. Kawai: *An Algorithm for Drawing General Undirected Graphs.* Information Processing Letters 31 (1989), North-Holland, 7–15.

28. R.M. Karp, R.E. Miller: *Parallel Program Schemata.* Journal of Computer and System Sciences, Vol. 3, 1969, 147–195.

29. C. Lin, D.C. Marinescu: *Stochastic High-level Petri Nets and Applications.* IEEE Transactions on Computers, 37 (1988), 815–825. Also in [26], 459–469.

30. M. Lindqvist: *Parameterized Reachability Trees for Predicate/Transition Nets.* In: G. Rozenberg (ed.): Advances in Petri Nets 1993, Lecture Notes in Computer Science Vol. 674, Springer-Verlag 1993, 301–324. Also in [26], 351–372.

31. G. Memmi, J. Vautherin: *Analysing Nets by the Invariant Method.* In: W. Brauer, W. Reisig, G. Rozenberg (eds.): Petri Nets: Central Models and Their Properties, Advances in Petri Nets 1986 Part I, Lecture Notes in Computer Science Vol. 254, Springer-Verlag 1987, 300–336. Also in [26], 247–283.

32. Y. Narahari: *On the Invariants of Coloured Petri Nets.* In: G. Rozenberg (ed.): Advances in Petri Nets 1985, Lecture Notes in Computer Science Vol. 222, Springer-Verlag 1986, 330–345.

33. M. Pedersen, J.B. Jørgensen, R.D. Andersen: *Occurrence Graphs with Equivalent Markings and Self-symmetries.* Master's Thesis, Computer Science Department, Aarhus University, Denmark, 1991. (Only available in Danish: *Tilstandsgrafer med Ækvivalente Mærkninger og Selvsymmetrier*).

34. L. Petrucci: *Combining Finkel's and Jensen's Reduction Techniques to Build Covering Trees for Coloured Nets.* Petri Net Newsletter no. 36 (August 1990), Special Interest Group on Petri Nets and Related System Models, Gesellschaft für Informatik (GI), Germany, 1990, 32–36.

35. *PNPM89: Petri Nets and Performance Models.* Proceedings of the 3rd International Workshop, Kyoto Japan 1989, IEEE Computer Society Press.

36. *PNPM91: Petri Nets and Performance Models.* Proceedings of the 4th International Workshop, Melbourne, Australia 1991, IEEE Computer Society Press.

37. *PNPM93: Petri Nets and Performance Models.* Proceedings of the 5th International Workshop, Toulouse, France 1993, IEEE Computer Society Press.

38. C. Reade: *Elements of Functional Programming.* International Computer Science Series, Addison-Wesley, 1989.

39. M. Silva, J. Martínez, P. Ladet, H. Alla: *Generalized Inverses and the Calculation of Symbolic Invariants for Coloured Petri Nets.* Technique et Science Informatiques 4 (1985), 113–126. Also in [26], 303–315.

40. M. Tiusanen: *Symbolic, Symmetry, and Stubborn Set Searches* In: R. Valette (ed.): Application and Theory of Petri Nets 1994. Proceedings of the 15th International Petri Net Conference, Zaragoza 1994, Lecture Notes in Computer Science Vol. 815, Springer-Verlag 1994, 511–530.

41. N. Treves: *A Comparative Study of Different Techniques for Semi-flows Computation in Place/Transition Nets.* In: G. Rozenberg (ed.): Advances in Petri Nets 1989, Lecture Notes in Computer Science Vol. 424, Springer-Verlag 1990, 433–452.

42. A. Valmari: *Stubborn Sets for Reduced State Space Generation.* In: G. Rozenberg (ed.): Advances in Petri Nets 1990, Lecture Notes in Computer Science Vol. 483, Springer-Verlag 1991, 491–515.

43. A. Valmari: *Stubborn Sets of Coloured Petri Nets.* Proceedings of the 12th International Conference on Application and Theory of Petri Nets, Aarhus 1991, 102–121.

Index

Monographs in Theoretical Computer Science – An EATCS Series

C. Calude
Information and Randomness
An Algorithmic Perspective

K. Jensen
Coloured Petri Nets
Basic Concepts, Analysis Methods and Practical Use, Vol. 1
2nd ed.

K. Jensen
Coloured Petri Nets
Basic Concepts, Analysis Methods and Practical Use, Vol. 2

K. Jensen
Coloured Petri Nets
Basic Concepts, Analysis Methods and Practical Use, Vol. 3

A. Nait Abdallah
The Logic of Partial Information

Texts in Theoretical Computer Science – An EATCS Series

J. L. Balcázar, J. Díaz, J. Gabarró
Structural Complexity I
2nd ed. (see also overleaf, Vol. 22)

M. Garzon
Models of Massive Parallelism
Analysis of Cellular Automata and Neural Networks

J. Hromkovič
Communication Complexity and Parallel Computing

A. Leitsch
The Resolution Calculus

A. Salomaa
Public-Key Cryptography
2nd ed.

K. Sikkel
Parsing Schemata
A Framework for Specification and Analysis of Parsing Algorithms

Former volumes appeared as EATCS Monographs on Theoretical Computer Science

Vol. 5: W. Kuich, A. Salomaa
Semirings, Automata, Languages

Vol. 6: H. Ehrig, B. Mahr
Fundamentals of Algebraic Specification 1
Equations and Initial Semantics

Vol. 7: F. Gécseg
Products of Automata

Vol. 8: F. Kröger
Temporal Logic of Programs

Vol. 9: K. Weihrauch
Computability

Vol. 10: H. Edelsbrunner
Algorithms in Combinatorial Geometry

Vol. 12: J. Berstel, C. Reutenauer
Rational Series and Their Languages

Vol. 13: E. Best, C. Fernández C.
Nonsequential Processes
A Petri Net View

Vol. 14: M. Jantzen
Confluent String Rewriting

Vol. 15: S. Sippu, E. Soisalon-Soininen
Parsing Theory
Volume I: Languages and Parsing

Vol. 16: P. Padawitz
Computing in Horn Clause Theories

Vol. 17: J. Paredaens, P. DeBra, M. Gyssens, D. Van Gucht
The Structure of the Relational Database Model

Vol. 18: J. Dassow, G. Páun
Regulated Rewriting in Formal Language Theory

Springer-Verlag and the Environment

We at Springer-Verlag firmly believe that an international science publisher has a special obligation to the environment, and our corporate policies consistently reflect this conviction.

We also expect our business partners – paper mills, printers, packaging manufacturers, etc. – to commit themselves to using environmentally friendly materials and production processes.

The paper in this book is made from low- or no-chlorine pulp and is acid free, in conformance with international standards for paper permanency.